Molded Optics

SERIES IN OPTICS AND OPTOELECTRONICS

Series Editors: **E Roy Pike**, Kings College, London, UK
Robert G W Brown, University of California, Irvine

Molded Optics
Design and Manufacture

Michael Schaub
Schaub Optical LLC, Tucson, Arizona

Jim Schwiegerling
University of Arizona, Tucson

Eric C. Fest
Phobos Optics LLC, Tucson, Arizona

Alan Symmons
LightPath Technologies, Orlando, Florida

R. Hamilton Shepard
FLIR Systems, Boston, Massachusetts

CRC Press
Taylor & Francis Group
Boca Raton London New York

CRC Press is an imprint of the
Taylor & Francis Group, an **informa** business

A TAYLOR & FRANCIS BOOK

CRC Press
Taylor & Francis Group
6000 Broken Sound Parkway NW, Suite 300
Boca Raton, FL 33487-2742

First issued in paperback 2020

© 2011 by Taylor and Francis Group, LLC
CRC Press is an imprint of Taylor & Francis Group, an Informa business

No claim to original U.S. Government works

ISBN-13: 978-0-367-57697-4 (pbk)
ISBN-13: 978-1-4398-3256-1 (hbk)

Visit the Taylor & Francis Web site at
http://www.taylorandfrancis.com

and the CRC Press Web site at
http://www.crcpress.com

The authors, in order of their biographical data, dedicate this book to

Elsa, Shadow, Shira, Patinhas, and Chaya

Diana, Max, Marie, and Mason

My wife, Gina, and my daughters, Fiona and Marlena

Lauren, Carter, Cooper, and Holden

Kelifern

Contents

Preface

Molded optics are currently being utilized in a wide variety of fields and applications. While in existence for almost one hundred years, advances in materials, machining capabilities, process control, and test equipment have spurred their increased use and acceptance in the past decade. The current desire for smaller, highly integrated, and more versatile products leads many engineers to consider them. Spanning the wavelength range from the visible to the infrared, they can be found in consumer electronics, medical devices, illumination systems, and military equipment, as well as a host of other products.

Molded optics provide designers with additional freedoms that can be used to reduce the cost, improve the performance, and expand the capabilities of the systems they develop. The use of aspheric and diffractive surfaces, which they lend themselves well to, has now become commonplace. The ability to reduce element count, integrate features, and provide for repeatable high-volume production will continue to keep molded optics in the trade space of many designs.

This book provides information on both the design and manufacture of molded optics. Based on the belief that an understanding of the manufacturing process is necessary to developing cost-effective, producible designs, manufacturing methods are described in extensive detail. Design guidelines, trade-offs, and best practices are also discussed, as is testing of several of their critical parameters. Additionally, two topics that often arise when designing with molded optics, mitigating stray light in systems employing them and mating such systems to the eye, are covered.

The authors, all experts in their particular areas, were selected based on both their knowledge and real-world experience, as well as their ability to transfer their understanding to others. I believe that they have succeeded in creating a work that will provide readers with information that will directly improve their ability to develop systems employing molded optics. Writing a text such as this takes considerable work, dedication, and time, and I thank the authors for their labors. Hours spent on the computer could well have been spent with family and friends, so it is no surprise that we dedicate this text to our spouses, children, and pets.

Mike Schaub
Tucson, Arizona

Authors

Michael Schaub
Schaub Optical LLC
Tucson, Arizona

Eric Fest
Phobos Optics LLC
Tucson, Arizona

Jim Schwiegerling
University of Arizona
Tucson, Arizona

Alan Symmons
LightPath Technologies, Inc.
Orlando, Florida

R. Hamilton Shepard
FLIR Systems
Boston, Massachusetts

1

Optical Design

Michael Schaub

CONTENTS

1.1 Introduction

The field of optical design can be considered a subset of the larger optical engineering discipline. Previously the domain of a relatively few, highly specialized individuals, optical design is now being performed by a wide range of persons, who may or may not have had a significant amount of optical training. This evolution in the field has resulted from a combination

1

of the availability of powerful personal computers, affordable optical design software, and the continually increasing use of optical technologies, including molded optics. In this chapter we cover some basics of optical design, highlighting important aspects as they relate to molded optics. We assume that the optical design work will be performed using one of the commercially available optical design software programs. We begin by discussing optical materials, then cover first-order and geometric optics, describe the types, effects, and control of aberrations, consider two special surfaces available to molded optics, and discuss methods of tolerance analysis and performance prediction.

1.2 Optical Materials

When discussing the optical properties of a material, two characteristics are normally specified: the refractive index of the material and the variation of its refractive index with wavelength. The refractive index of the material, whose value should be quoted with respect to a particular wavelength, is the ratio of the speed of light in a vacuum to the speed of light in the optical material. Thus, the higher the index of refraction, the slower light travels in the material. As we shall see later, materials with higher refractive indices refract (bend) light more than materials with lower refractive indices.

The variation of the refractive index with wavelength, known as the dispersion of the material, is usually specified by a single value called the Abbe or *V* number. For the visible spectrum, the Abbe number is defined as

$$V = \frac{n_d - 1}{n_F - n_C} \tag{1.1}$$

where n_F, n_C, and n_d refer to the refractive index of the material at wavelengths of 486.1, 656.3, and 587.6 nm, respectively. The lower the Abbe number of the material, the more dispersive it is, and the more the refractive index of the material varies with wavelength. Abbe numbers can also be defined for other spectral regions, such as the short-wave infrared (SWIR), as will be discussed in Chapter 6 on molded infrared optics. Materials with Abbe numbers below about fifty are referred to as flints, while materials with Abbe numbers above fifty are referred to as crowns. The dividing line between crowns and flints is somewhat arbitrary. Using the term *crown* or *flint* to describe a material alerts the designer to its relative dispersion. Because of the importance of these two quantities, index and dispersion, optical glasses are sometimes specified using a six-digit code that indicates their values. For instance, N-BK7, a common optical glass, has an index of refraction (at 587.6 nm) of 1.517 and a dispersion (using n_F, n_C, and n_d) of 64.2.

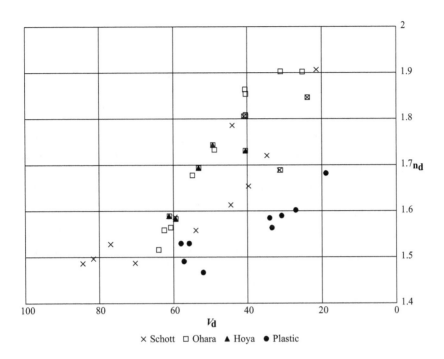

FIGURE 1.1
Glass map showing common visible molded glass and plastic optic materials.

Thus, the glass may be specified as 517642, where the first three digits correspond to the glass's index (the remainder of the index minus one, 1.517 − 1 = 0.517) and the last three digits correspond to the dispersion (the dispersion times ten, 64.2 × 10 = 642). Specifying glasses in this manner allows comparison of substitute glasses from across multiple glass vendors. However, these two numbers do not encompass all properties of the glass, and care should be exercised when substituting glasses in a developed design.

In order to show the optical material choices available to the designer, it has become common for material manufacturers to create a plot with each glass represented as a point on a graph having its axes as index and dispersion. Such a plot is referred to either as a glass map or an n-V diagram, a version of which is shown in Figure 1.1, displaying a variety of moldable glass and plastic materials for the visible region. The refractive index of each material is plotted as the ordinate and the dispersion plotted as the abscissa. Note that the values along the horizontal (dispersion) axis are plotted in reverse. The data used to create this map were taken from several vendors and contain a representative sampling of commonly available moldable optical materials.

Several things are readily apparent from the glass map. First, there is a wide range of refractive index and dispersion choices available for moldable glass materials. In general, the higher the refractive index of the glass, the more dispersive it is, though there is not a strict relation between the two.

Second, the optical plastics tend to have lower refractive indices than glasses of comparable dispersions. Third, there are fewer plastic optical materials available than the moldable optical glasses.

In addition to refractive index and dispersion, there are a number of other properties to consider when selecting an optical material. In general, we want the material to be highly transmissive in the spectral region we are using. We would like the material to have known, consistent properties for its change with temperature, in particular the change of index with temperature, known as dn/dt, and its expansion with temperature, known as its coefficient of thermal expansion (CTE). We want the material to be appropriately machinable, workable, and in the case of molded optics, relatively easily moldable. We also want the material to have suitable environmental resistance to the conditions it will see, whether they be thermal environments, vibration, moisture, abrasion, or chemicals. Cost, availability, resistance to staining, the ability to be coated, and lifetime stability are additional factors to consider. In many cases, all of the desired properties cannot be simultaneously met, leaving it to the designer to make the selection of the best optical material available.

1.3 Geometric Optics

Geometric optics is concerned with the propagation of light based on the concept of rays. Interpreting Fermat's principle, light will pass through a system along a path that takes the minimum time (the path is actually mathematically stationary, but this is often the minimum). Rays are considered the paths that the light will follow, which are straight line segments within a homogeneous material. Light traveling in straight lines is familiar to anyone who has witnessed their shadow or observed a beam of light from a hole in a window covering traversing a slightly dusty room. In reality, light does not travel in perfectly straight lines, but diffracts (spreads out) as it passes through apertures. For most purposes of optical design, however, the ray approximation is sufficient to accurately calculate the passage of light through an optical system. We begin our discussion of geometric optics with the subject of first-order models, which allow us to evaluate the location and size of images created by an optical system.

1.3.1 First-Order Optics

A first-order optical model is a description of an entire optical system by a set of six points along the optical axis, known as the cardinal points. The cardinal points consist of the front and rear principal points (P, P'), the front and rear nodal points (N, N'), and the front and rear focal points (F, F'). As an

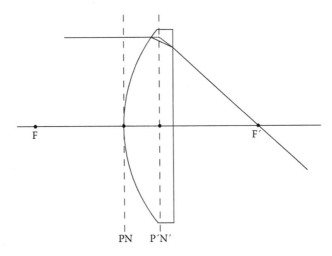

FIGURE 1.2
Cardinal points of a convex-plano lens.

example, the cardinal points for a convex-plano lens are shown in Figure 1.2. Also included in the figure is a ray passing through the lens. On the left side of the figure the ray is parallel to the symmetry axis of the lens.

If planes are drawn perpendicular to the axis at the principal points, they are referred to as the principal planes, which are shown as the dashed vertical lines in the figure. The principal planes can be considered the effective locations of ray bending within the system. Even though the actual bending of the rays occurs at the surfaces of the lens, from an input/output aspect, the rays appear to bend at the principal planes. Rays input to the front of the system, parallel to the axis, appear to bend at the rear principal plane. Thus, if we extend a ray segment entering the system parallel to the axis and the corresponding ray segment exiting the system (as seen by the dashed extensions in the figure), they intersect at the rear principal plane. The principal planes (and points) are the planes (and points) of unit transverse magnification, in that a ray striking one principal plane is transferred to the other principal plane at the same height. This is true whether or not the input rays are parallel to the axis. Thus, a general ray entering the system will bend at the front principal plane, be transferred at the same height to the rear principal plane, bend at the rear principle plane, and exit the system.

While the principal points are points of unit transverse magnification, the nodal points are the points of unit angular magnification. Thus, if a ray is input to the system heading toward the front nodal point, it will exit the system appearing to emerge from the rear nodal point, with the same angle to the axis as the input ray. When the system has the same refractive index in front of and behind it, as is most often the situation, the nodal points and principal points are coincident. Unlike the case for the principal points, there are no nodal planes. Unity angular magnification only occurs for rays aimed toward the nodal points themselves.

The focal points occur at the axial location where rays, input parallel to the system axis, cross the axis after passing through the system. Such a ray, input to the front of the system, crosses the axis at the rear focal point, while a ray parallel to the axis that enters the rear of the system crosses the axis at the front focal point. Another way of viewing this is that rays entering the system that pass through the front focal point exit the system parallel to the axis. The rear focal length of the system is defined as the distance from the rear principal point to the rear focal point. The front focal length of the system is defined similarly. The power of an optical surface, which is the reciprocal of the surface's focal length, is directly related to the index of the lens material and inversely proportional to the surface's radius of curvature. Thus, for a given index, shorter radii surfaces have more optical power than longer radii surfaces. The power of the entire lens, which is the reciprocal of the focal length of the lens, depends upon the power of each surface and the distance between them.

The focal length of the system acts as a scaling factor for its first-order imaging properties. For instance, if the same distant object is viewed with two lenses, one with a focal length twice that of the other, the image formed by the longer focal length lens will be twice the size of that by the shorter focal length lens. Of course, if the image is captured in each case by the same size detector, the longer focal length lens will provide only half the view angle captured with the shorter focal length lens. The back focal distance of a system is the length from the last optical surface of the system to the rear focal point. This distance, which should not be confused with the focal length of the system, can be an important parameter in a design if space is needed behind the optical system for items such as fold mirrors or beamsplitters.

With a first-order optical model defined, we can compute the image location and size of any input object. For systems with the same refractive index on both sides, the image location is related to the object location through the equation

$$xx' = -ff' \tag{1.2}$$

where x is the distance from the front focal point to the object, x' is the distance from the rear focal point to the image, and f and f' are the front and rear focal lengths, respectively, of the system. An example of such an imaging arrangement is shown in Figure 1.3. Distances are measured from the focal points, moving to the left being negative and moving to the right positive. In the figure, the distance x is negative, while x' is positive.

To determine the size of the image, we can use the following equation:

$$h' = \frac{fh}{x} = -\frac{x'h}{f} \tag{1.3}$$

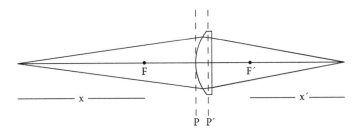

FIGURE 1.3
Imaging setup using a convex-plano lens.

where h is the object height, h' is the image height, and x, x', and f have the same meanings as above, and we have assumed f and f' are equal. The ratio of the image height to the object height is the magnification of the system.

For the case of an infinite object distance, direct application of the equations would yield an image location of $x' = 0$ as well as an image height of $h' = 0$. This would indicate that the image is located exactly at the focal point of the system. In reality, the object is at some finite distance, so the image would have some finite height and be located slightly off the focal point.

The location of the cardinal points for an optical system can be calculated using knowledge of the physical parameters of the elements contained in it. The values needed for the calculation are the radii of curvature of the optical surfaces, their locations to one another, and the refractive index of the optical materials surrounding them. We do not discuss the details of this calculation here, but refer the reader to several references.[1–4]

1.3.2 Pupils and Stops

Up to this point, we have not concerned ourselves with the size of the optical elements or the beams passing through them. However, if we intend to design an actual system, we need to consider whether a given ray makes it through the system or not. This leads us to the concept of pupils and stops. There are two kinds of stops, the aperture stop and the field stop, as well as two pupils, the entrance pupil and exit pupil, in each optical system.

The stops, as their names imply, limit the size of the aperture and field angle of the system. In every optical system, there is some aperture that limits the size of the on-axis beam that can pass through the system. This aperture may be the diameter of a lens element, the edge of a flange that a lens is mounted on, or an aperture placed in the system (such as the iris in a digital camera) to intentionally limit the beam size. Whatever it may be, the aperture that limits the size of the on-axis beam through the system is known as the aperture stop, which is often referred to simply as "the stop." The aperture stop may be buried inside the system or may be in front of or behind all the optical elements. The field stop sets a limit on the field of view of the system. That is, it sets an upper limit on the angular input of beams that form the captured image. In many cases, the field stop is the image capture device of the system itself. In digital cameras, where the image is captured by a detector, any light that falls outside the edges of the detector is not (intentionally) collected. Thus, the detector is limiting how big a field is seen by the system, making it the field stop. The term *field stop* is also used to describe slightly oversized apertures placed at or near intermediate images within the system. While not strictly field stops by the definition above, since they do not actually limit the field of view, these apertures can help to control stray light.

The pupils can be considered the windows into and out of the system. The pupils are simply the images of the aperture stop, viewed through all the optics between the viewer and the stop itself. Looking into the back of a camera lens, we can see an effective aperture from which all the light appears to come. This effective aperture is known as the exit pupil. Since all beams pass through the aperture stop, all beams appear to pass through its image, the exit pupil. Similarly, looking into the front of the camera lens, we see the entrance pupil, through which all beams appear to enter the system. If the aperture stop is in front of all the optical elements, the entrance pupil is located at the same position as the aperture stop (since there are no elements for the stop to be imaged through). Similarly, if the aperture stop is behind all the optical elements, the exit pupil and aperture stop are coincident.

The ratio of the focal length of a system to the diameter of its entrance pupil is an important quantity known as the F/number or F/stop, usually shown as F/#. This ratio is important in that it relates to the amount of light captured by the system, through the solid angle, as will be discussed in Chapter 3. The lower the F/# of the system, the more relative light gathering ability it has. Systems with lower F/#s, F/2 for example, are said to be faster than systems with larger F/#s, such as F/10. This term is historical in that a faster lens (lower F/#), being better at light gathering, requires the shutter on a camera to be open for less time than a slower (higher F/#) lens. Thus, the picture is taken more quickly with a "faster" lens.

1.3.3 Snell's Law and Ray Tracing

While first-order optics allows us to calculate the location and size of an image, a critical aspect of optical design that it does not provide is a prediction of the quality of the image. By image quality, we mean how well the image represents the true characteristics of the object, such as its fine detail. To predict the image quality provided by an optical system, we turn to ray tracing.

The fundamental rule used in ray tracing is Snell's law. This law relates the direction of a ray after an optical surface to the direction of the ray before it through the ratio of the refractive indices of the materials on each side of the surface.

Snell's law is shown below:

$$n \sin \theta = n' \sin \theta' \qquad (1.4)$$

where n is the index of refraction of the material before the surface, n' is the refractive index of the material after the surface, and θ and θ' are the angles of incidence and refraction. These angles are measured with respect to the normal to the surface at the point of intersection of the ray. It should be noted that the incident ray, the refracted ray, and the surface normal all lie in a common plane.

In the case of mirrors, the index of refraction after the mirror is taken as $n = -1$. In this special case, we see that the magnitude of angle of incidence and refraction (actually reflection) are equal to each other. The reflected ray lies on the opposite side of the normal to the surface, at the same angle as the incident ray. As above, the incident ray, reflected (refracted) ray, and surface normal all lie in the same plane.

We previously stated that materials with higher refractive index bend light more than materials with lower refractive index. This can be shown by direct substitution into Snell's law. Consider a ray that is incident on the boundary between air ($n = 1$) and a planar glass surface at an angle of incidence (θ) of 45°. If the glass has an index (n') of 1.5, the angle of refraction (θ') would be 28.13°, while if the index of the glass is 1.7, the angle of refraction would be 24.58°. Since the angle of refraction is measured with respect to the surface normal, the ray for the refractive index of 1.7 is closer to the normal than the ray for the refractive index of 1.5, and thus has been deviated further from its original direction of 45° to the normal. Therefore, we see that the higher-index material bends the ray more from its original direction than the lower-index material does.

If the surface where refraction occurs is not planar, but another shape, such as spherical, we can still apply Snell's law. In the same manner as before, we determine the angle of incidence with respect to the normal to the surface (which for a spherical surface is its radius), substitute the appropriate values into Snell's law, and determine the angle of refraction. Successive application of Snell's law, along with the transfer of rays between optical surfaces, allows us to determine the paths of rays through an entire optical system. As a simple example, consider the ray passing through the section of a convex-plano lens shown in Figure 1.4. The input ray, parallel to the axis, strikes the lens at a height of 1.75. At this location, the ray makes an angle of 35.685° with the normal to the surface. Applying Snell's law, we determine that the angle of refraction is 22.59°. Using the refracted angle, the direction of the normal, and the distance to the planar surface, we determine that the ray strikes the rear surface of the lens at a height of 1.6484 and an angle of incidence of 13.09°. Again applying Snell's law, we determine the refraction angle to be 20.123°. Using this angle, the direction of the surface normal (parallel to

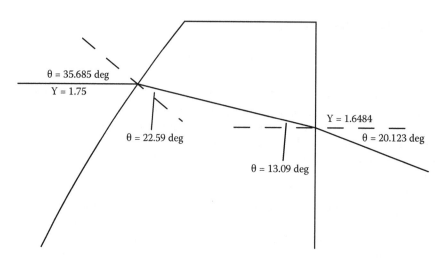

FIGURE 1.4
Ray trace through a convex-plano lens.

the axis in this case), and the height of the ray at the rear surface, we can determine where the ray crosses the axis by solving for a ray height of zero. Solving, we determine a ray height of zero occurs at a distance of 4.4989 from the rear surface of the lens. We have now traced a single ray through our one-element optical system. Ray tracing calculations were previously performed manually, using log tables or basic calculators, depending on the era. Needless to say, tracing even a single ray through a multielement system took considerable effort. With modern computers, thousands of rays can be traced per second, greatly simplifying the calculations required of the optical designer.

If we were to trace multiple rays in this input beam, entering the lens at varying heights, we would find that all the rays do not pass through the same axial point, as would be predicted by first-order optics. This is due to the presence of aberrations, which are not included in our first-order model and are the subject of the next section.

1.4 Aberrations

Aberrations are deviations from the perfect imagery predicted by first-order optics. We can consider aberrations to fall into one of two general categories: those that result in a point not being imaged to a point, and those that result in a point being imaged to a point in the incorrect location. Examples of the first type are spherical aberration, coma, and astigmatism, while examples of the second type are Petzval curvature and distortion.

Aberrations are a direct result of the nonlinearity of Snell's law, along with an optical surface's shape and location within the optical system. The surface's shape, as well as its distance from the aperture stop, determines the angles of incidence for the rays striking it. By adjusting the angles of incidence, we can control how large or small the surface's aberration contribution is. The amount of each aberration also varies as a function of the entrance pupil diameter or the field angle of the system. We now briefly discuss each of the aberrations mentioned, along with methods of controlling them.

1.4.1 Spherical Aberration

Spherical aberration is the variation in focus position of a ray as a function of its radial distance from the center of the entrance pupil. This is illustrated in Figure 1.5, which shows a collimated input beam passing through a convex-plano lens. The rays striking the outer portion of the lens cross the axis closer to the rear surface than rays striking the lens near its center. The lens surfaces, spherical at the front and planar at the rear, refract each of the rays in the input beam according to Snell's law. For this lens, the rays near the edge of the lens are refracted more sharply than is needed for them to come to a focus position coincident with the rays in the center, a condition referred to as undercorrected spherical aberration.

The magnitude of the spherical aberration is measured as the lateral distance, at the image plane, between a ray from the edge and the center of the pupil. The amount of spherical aberration depends upon the third power of the pupil diameter, but is independent of the field angle. Thus, operating a system at twice its original pupil diameter will result in an eightfold increase in spherical aberration, with the aberration being a constant value over the field of view. Conversely, reducing the aperture stop size (and thus the pupil size), known as stopping down the system, will reduce the spherical aberration.

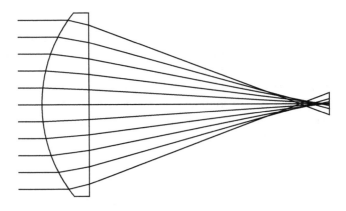

FIGURE 1.5
Convex-plano lens exhibiting spherical aberration.

Control of spherical aberration can be achieved by several methods. The most common method is "bending" the lens, which is adjusting the radii of the two surfaces while maintaining the overall power of the element. In the case of the lens shown in Figure 1.5, reducing the spherical aberration by this method would result in a longer radius of curvature for the front surface and a convex radius (positive power) for the second surface. By increasing the radius of the front surface, the angles of incidence of the rays are decreased, reducing the spherical aberration contribution of the surface. Adding power to the rear surface maintains the focal length of the element, but increases the angles of incidence on it and its spherical aberration contribution. However, the magnitude of the increase at the rear surface is not as great as the decrease in the front surface contribution, resulting in an overall reduction in the spherical aberration of the element.

Another method of reducing spherical aberration is to "split" the element into two elements whose combined power is the same as that of the original lens. Splitting the lens allows longer radii of curvature for each of the surfaces, reducing the angles of incidence of the rays, and decreasing the associated surface spherical aberration contributions. Of course, splitting a lens requires the introduction of an additional element, which may have negative impacts on cost, space, and weight.

A third method of reducing spherical aberration is to fabricate the element from a higher refractive index material. Selection of a higher-index material allows longer radii for a given power, which again reduces the angles of incidence and the associated spherical aberration contribution. For molded plastic optics, selection of a higher-index material may not be possible due to the limited material choices. For molded glass optics, higher refractive index materials are generally available. It should be noted that higher refractive index glasses tend to be more expensive, as well as denser, than lower-index materials. This trade on cost and weight needs to be evaluated during the design process.

The final method of controlling spherical aberration we discuss is the use of an aspheric surface. Unlike spherical surfaces, which have a single curvature based on their radius, the curvature of an aspheric surface can be defined as a function of surface position. This allows, within reason, the surface to be tailored to provide the correct angles of incidence to eliminate spherical aberration. As we shall see later, changing from a spherical to an aspheric surface for the lens of Figure 1.5 allows removal of the spherical aberration of the element.

1.4.2 Coma

Coma is a variation in magnification as a function of ray position in the pupil. We previously defined magnification as the ratio of the image height to the object height. In the presence of coma, rays from a given point on the object

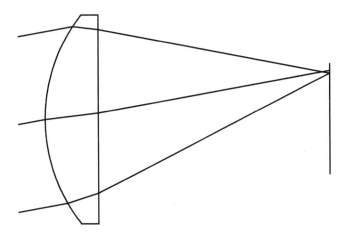

FIGURE 1.6
Convex-plano lens exhibiting coma.

enter the pupil at various locations and then produce image heights of different values. Thus, the point on the object is not imaged to a point, but to a blur whose size is related to the variation in the magnification. An example of coma is shown in Figure 1.6, where rays from the edges and center of the pupil, which is located at the front lens surface, strike the image plane at differing heights. In this figure we have moved the image plane to the location where the rays at the edge of the pupil intersect. Coma has the interesting property that rays from annular rings in the pupil form circles on the image plane. The larger the diameter of the annular pupil ring, the larger the image plane circle, resulting in a series of circles at the image plane that ultimately form a comet-like image, which is the basis for the name of the aberration.

Coma depends upon the square of the diameter of the pupil and is linearly dependent on the field. Thus, the image degradation grows as we move away from the axis of the system. Coma can be controlled in several ways. The first method is the use of symmetry. If the optical system is symmetric about the aperture stop, the contributions of coma from the elements in front of the stop will be opposite to those from the elements behind the stop. Making a system perfectly symmetric about the stop is not common, but a high degree of symmetry is often seen in systems such as camera lenses. Another method of controlling coma is to adjust the location of the aperture stop relative to the lens elements. Changing the location of the aperture stop changes the location on the lenses that beams forming the off-axis image points strike. By adjusting the location of the beams on a surface, we change the angles of incidence of the beam, which changes the surface's aberration contribution. A final method of controlling coma is the use of aspheric surfaces, which again allow adjustment of the surface as a function of its height, and therefore adjustment of the angles of incidence and amount of aberration introduced.

1.4.3 Astigmatism

Astigmatism is the variation in focus for rays in two orthogonal planes through the optical system, as shown in Figure 1.7. One plane, known as the meridional plane, is a plane of bilateral symmetry of the system and is the plane of the drawing in Figure 1.6, containing rays from the top and bottom of the pupil. The rays for this plane are shown in the top of Figure 1.7. The other plane, the sagittal plane, contains the rays from the left and right of the pupil and is perpendicular to the drawing in Figure 1.6. These rays are shown in the bottom of Figure 1.7, which is a top view of the lens. At each of the two focus locations, meridional and sagittal, the rays form a line segment, due to the set from one plane being in focus and the set from the other plane being out of focus. In between these two focus positions the rays form an elliptical spot, with the ellipse degenerating to a circular spot midway between. The image plane in the figure has been positioned midway between the focus of the meridional and sagittal rays.

Astigmatism depends linearly on the pupil diameter and with the square of the field. This aberration arises because the meridional and sagittal beam

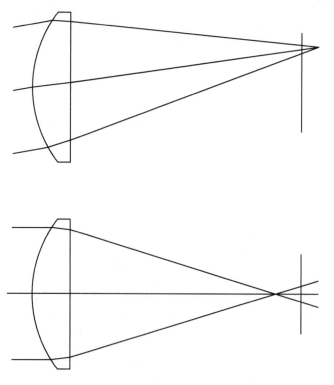

FIGURE 1.7

Astigmatism, with meridional rays shown in side view (above) and sagittal rays shown in top view (below).

widths are different for a beam striking a surface off axis. The two beam widths thus see different angles of incidence and are refracted by different amounts. Astigmatism can be controlled by adjusting the shape of the surfaces, as well as their location in relation to the aperture stop of the system. It can also be controlled through the use of aspheric surfaces.

1.4.4 Petzval Curvature

Petzval curvature refers to the curvature of the ideal image surface for a powered element, as shown in Figure 1.8. For a positive lens the ideal image surface curves inward toward the lens, while the opposite is true for a negative lens. The amount of Petzval curvature depends linearly on the power of the element and inversely on its refractive index. Petzval curvature, unlike the aberrations previously discussed, does not directly blur the image of a point. The image of a point object in the presence of Petzval curvature is still a point, provided that the image is evaluated on a curved image surface. If we were to evaluate the image of point sources at varying heights on a planar image surface coincident with the on-axis focus location, we would find that the points are increasingly blurred as we move to larger image heights. This results from the evaluation taking place further and further from the ideal curved image surface. In most imaging applications we do not use a curved image surface, but instead use a flat surface such as a detector. This requires us to control the amount of Petzval curvature in order to obtain an acceptable image. As Petzval curvature is related to the power and index of each optical element, the control of this aberration can be achieved by using a combination of elements of both positive and negative powers, along with proper selection of their refractive index values.

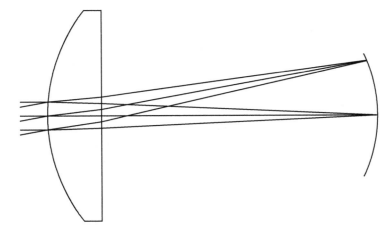

FIGURE 1.8
Ideal image surface for a positive lens, showing Petzval curvature.

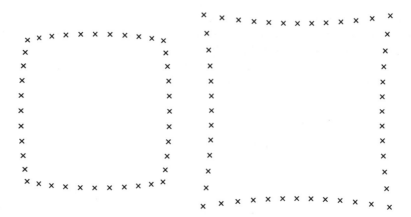

FIGURE 1.9
Images of points on a square for barrel (left) and pincushion (right) distortion.

1.4.5 Distortion

Distortion is likely the aberration that most people are familiar with. The effect of distortion can easily be seen in images produced by very wide-angle lenses. In these images, straight lines in the object are increasingly curved the farther out they are from the center of the image. Distortion does not blur the image of a point. Instead, distortion changes the location of a point's image relative to the location predicted by first-order optics. The amount of distortion is usually quoted as a percentage, as a function of the field of view. The percent distortion is calculated as the difference between the real and first-order image heights, divided by the first-order image height. The percentage of distortion varies as the square of the field.

The effect of distortion, as mentioned above, is the curving of straight lines. The direction of curving, inward or outward, depends on the sign of the distortion. Negative distortion curves the edges of the lines inward relative to their center, resulting in a barrel shape, while positive distortion curves the edges of the lines outward, resulting in a pincushion shape. These shapes, the distorted images of points on a square object, are displayed in Figure 1.9, where the magnitude of the distortion at the edges of the images is about 15%. Like coma, distortion can be controlled through the use of symmetry about the stop. The distortion produced by an element is related to its distance from the aperture stop, so adjustment of this relationship can also be used to control distortion. Additionally, aspheric surfaces can be used in the control of distortion.

1.4.6 Axial Color

There are two basic aberrations related to the dispersion of optical materials. These are axial color, discussed here, and lateral color, discussed in the next section. These chromatic aberrations, as they are called, can be thought of as

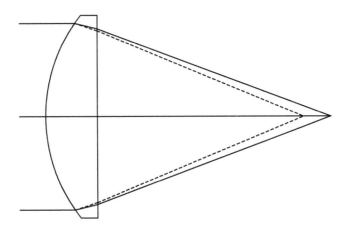

FIGURE 1.10
Exaggerated depiction of axial color, with blue (dashed) and red (solid) rays shown.

variations in the first-order properties of the optical system as a function of wavelength. We previously discussed that the focal length (or inversely power) of an optical element depends in part on the refractive index of the material it is made of. We have also discussed the dispersion of optical materials, that is, their change in refractive index with wavelength. Since the focal length depends on the refractive index and the refractive index changes value with wavelength, it makes sense that the focal length would change value with wavelength.

The result of this change with wavelength is axial color (sometimes called longitudinal color), which is the condition where different colors (wavelengths) come to focus at different points along the axis of the system. An exaggerated example of this is displayed in Figure 1.10, where the dashed rays after the front lens surface represent blue light and the solid rays after the front lens surface represent red light. In general, the refractive index of optical materials gets higher for shorter wavelengths (toward the blue end of the visible) and decreases for longer wavelengths (toward the red end of the visible). This results in a shorter focal length for blue light than for red light. Thus, the blue light focuses closer to the lens than the red light, as seen in the figure. The amount of longitudinal (axial) separation of the blue and red light depends inversely upon the dispersion of the lens material (the V number) and directly with the focal length of the lens.

Control of axial color can be achieved by combining elements of positive and negative power. A negative element has the opposite sign of axial color from that of a positive element. An example of a simple optical system that is corrected for axial color is an achromatic doublet. This system consists of two lenses, one positive and one negative, whose axial color contributions cancel each other. Because the axial color contribution of a lens depends on both its power and material dispersion, there are a number of combinations of lens powers and materials that can be paired in the doublet to correct axial color. The basic equation for achieving axial color correction with two lenses (thin lenses in contact with each other) is

$$\frac{\phi_1}{V_1} + \frac{\phi_2}{V_2} = 0 \qquad (1.5)$$

where ϕ_1 is the power of the first element, ϕ_2 is the power of the second element, and V_1 and V_2 are the dispersions of the two lens materials.

The simplest case of correcting axial color would consist of making two lenses of the same material, of equal and opposite power. This would provide axial color correction, but the powers of the two lenses would also cancel, providing no focusing. The doublet would essentially be a window. Thus, in order to make a doublet with a finite focal length, the two lenses are made of different materials and with different powers. Typically, the positive lens has more power and is made of a low-dispersion (high V number) material, while the negative lens has less power and is made of a high-dispersion (low V number) material. The difference in powers results in an overall positive system power, while the sum of the ratios of the powers and dispersions equals zero, resulting in correction of axial color. The use of different dispersion materials in this example points to the desire of designers to have a variety of materials available to them.

We should note that correction of axial color refers to the fact that the red and blue wavelengths come together at the same focus. The green wavelength, however, is not coincident with the red and blue. This departure of the green wavelength from the red/blue focus is known as secondary color. This can be significantly more difficult to control, but typically has a much smaller value than the original red and blue separation.

1.4.7 Lateral Color

While axial color changes the focus position with wavelength, lateral color changes the image height as a function of wavelength. Lateral color, also referred to as transverse color, increases linearly with field, making it difficult to control in systems with a wide field of view. Lateral color can be observed in an image as color fringing at the edge of objects, due to the mismatch in heights between the various wavelengths. The amount of lateral color that an element introduces is related to its distance from the aperture stop. Lateral color, like coma and distortion, can be controlled through the use of symmetry. It also can be controlled by achromatizing elements (for instance, by making them doublets) or by balancing the contributions of the various elements within the system.

1.4.8 Chromatic Variation of Aberrations

While we have just discussed the main chromatic aberrations, axial and lateral color, each of the aberrations discussed before them, spherical, coma, etc., can also vary with color. For example, the variation in spherical aberration with wavelength is known as spherochromatism. This tends to be more of a second-order effect, but one that the designer must still be aware of. It would not make sense to go to the effort of correcting spherical aberration at the center of the wavelength band, but allow large changes in spherical aberration with color to degrade the image.

The amount of aberration change produced by a surface over the wavelength band is approximately equal to the amount of aberration produced at the center of the band divided by the dispersion (V number over the band) of the material the surface is made of. It is common when designing with molded optics to utilize aspheric surfaces, which are discussed in Section 1.6.1. Using these, it is possible for a single surface to create large amounts of (desired) aberration, which can also introduce large amounts of chromatic variation of the aberration.

1.5 Optical Design with Molded Optics

Overall, the design of optical systems using molded optics is similar to the design of standard optical systems. Today, most optical design is performed on personal computers, using commercially available optical design and analysis software programs. The software is readily equipped to handle the design of systems employing molded optics. Molded optics provide additional variables, such as aspheric or diffractive surface coefficients, which can be used to optimize the performance of a design.

In general, optical design consists of developing a system that has the desired first-order optical properties and adequate performance (when built). Additionally, the design should be able to be cost-effectively manufactured in the volumes needed. For imaging systems, performance is usually tied to an image quality metric such as ensquared energy, modulation transfer function (MTF), or wavefront error, while illumination systems or other applications will have their own specific performance metrics. Broadly stated, imaging systems with any or all of the parameters of faster F/#s, larger fields, and wider wavebands make it more difficult to achieve a given performance metric value. For a given F/#, field of view, and waveband, increases in performance metric value are typically obtained by increasing the complexity of the design. This increase in complexity may result from using any, several, or all of a larger number of elements, less common materials, increasingly tight tolerances, or alternate surface forms, such as aspheres. Figures 1.11 to 1.13 show an example of performance change with complexity increase, using simulated images created with commercial optical design software. The input to the simulation is shown in Figure 1.11. The image used for the input was captured with a commercial digital camera (Nikon D90), using an

FIGURE 1.11 (See color insert.)
Input (object) for image simulation. (Courtesy of Dr. Eric Fest.)

F/2.8, 24 mm focal length lens. To show the performance change with system complexity, a single-element lens and a three-element lens are compared. Figure 1.12 shows the simulated image for the single-lens system, while Figure 1.13 shows the simulated image for the triplet. Both systems are operating at F/4.8, with a focal length of 24 mm. Note that this is a slower F/# than was used for taking the input image. The image quality difference is obvious between the one- and three-element systems. As expected, having a larger number of elements improves the image quality (assuming both systems can be comparably built). The single element system provides little opportunity (few variables) for aberration control across the field of view and is not color corrected. In contrast, the three-element system provides additional variables (surface radii, lens spacings, use of multiple optical materials), which allows greater aberration control. Even with these extra parameters the triplet does not provide enough aberration control for full field coverage, as blurring can be seen at the edges and corners of the image. In comparison, the lens used to capture the input image is operating at a faster F/# and provides excellent image quality over the entire field. Its complexity, however, is much greater, as it is composed of nine lens elements.

For a given number of elements and a defined waveband, the trade between F/# and field of view is fairly straightforward; in order to achieve a fixed performance metric value, larger fields of view require slower F/#s and faster systems require smaller fields. When using molded optics, cost may limit the number of elements that the designer can use. This may set the maximum

FIGURE 1.12 (See color insert.)
Simulated image using a single-element system.

FIGURE 1.13 (See color insert.)
Simulated image using a three-element (triplet) system.

performance that can be obtained for a given F/# and field. While the use of aspheric and diffractive surfaces can improve system performance, it may also be that their addition to the system provides less performance increase than would be obtained by simply adding another element. The trade between performance and number of elements is a common one, particularly for systems employing molded optics. The final decision on the trade is often based on cost, with customers wanting better performance, but not wanting (or being able) to pay for it, due to their cost model and product markets.

While the devil is in the details, the optical design process itself is relatively straightforward, as is described elsewhere in much greater detail.[5-9] With some starting point for the system, whether a previous design or merely slabs of glass or plastic, a performance metric known as a merit function is developed. The merit function describes the system performance, for example, spot size, by evaluating where rays from a given point source cross the image. It typically contains constraints on first-order properties of the system such as focal length. The merit function may also contain additional user-entered inputs, such as constraints on edge thickness for molded plastic optics. The optimization algorithm of the software seeks to minimize the merit function by adjusting parameters of the system, such as surface radii, which have been defined by the designer as available variables. Being a multidimensional problem, there may be various local minima within the space defined by the merit function. Part of the role of the designer is to determine if the optimization algorithm has stagnated in a local minimum or if the optimum solution, as defined by the merit function, has been reached. Most optical design programs currently have features to search for the global minimum, that is, the best solution. It should be kept firmly in mind that the best solution is defined by the constraints that have been placed (or not placed) on the system. For instance, it may be that a design with outstanding performance is easily found, but is also not manufacturable. Again, this is where the designer needs to play an active role in development of the system.

Understanding the manufacturing methods of molded optics and the constraints that these methods place on the molded elements themselves is a key to successfully designing systems employing them. Specific manufacturing methods and constraints are discussed in later chapters. In addition to understanding the constraints, understanding how to implement them in the optical design software is important, as is an understanding of how this implementation affects the optimization process. Entering a specific constraint is software-selection dependent, though there are typically several ways of adding any particular constraint to the merit function. Additionally, the weight, or importance of the constraint, can also be set and adjusted. Weighting a constraint too heavily may reduce the number of viable systems the optimization algorithm can develop, while lightly weighted constraints may end up not constraining the parameter adequately. The designer should look at different constraints and weightings during the design process to verify that they are not overly burdening the system. At the same time, he or she needs to ensure the components are properly constrained so that they are manufacturable.

Earlier, we discussed the use of ray tracing to evaluate the image quality of a system. In addition to image degradation due to rays from a point object not all coming to focus at the same point, diffraction also affects final image quality. Diffraction, which is related to the wave nature of light, ultimately limits the size of the spot to which a beam can be focused. Thus, while a spot diagram or ray fan within the optical design software may indicate that the image of a point object is a true point, in reality there is a finite physical size that the point image must be above. An example of a "perfect" point image, including the effects of diffraction, is shown in Figure 1.14, along with a cross section through its center. The perfect system used to model this is operating at F/2.8, with a focal length of 24 mm and light of wavelength

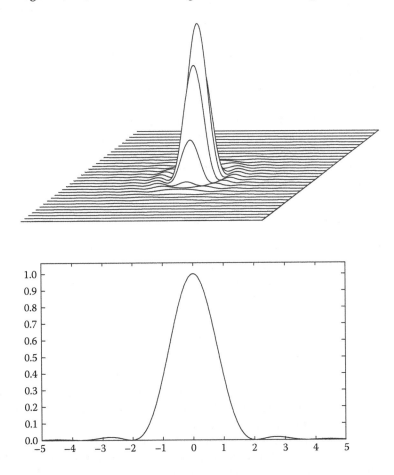

FIGURE 1.14
Image of a point object through a "perfect" system, showing effects of diffraction (Airy pattern). Horizontal scale on cross section is in microns.

587.6 nm. This point image, limited only by diffraction, is known as an Airy pattern or Airy disk. Final image quality calculations performed by optical analysis software include the effect of diffraction from the appropriate aperture in the optical system, unless told to do otherwise.

1.6 Optical Surfaces

Optical surfaces have traditionally been spherical in shape, as this has been the easiest surface form to create, specify, and test. Random motion between an abrasive-covered surface and a piece of glass forms a smooth, spherical surface on the glass. Specifying a spherical surface, at least its nominal form, requires only the value of the radius of curvature. Spherical surfaces can even be tested with other spherical surfaces, known as test plates, by examining the light pattern formed when the surface and test plate are brought into close proximity.

Because molded optical surfaces are formed using replication from a master surface within a mold, they do not need to be spherical in order to be easily produced. In fact, in many cases it is as easy to produce a molded nonspherical surface as it is to produce a molded spherical one. As a result, nonspherical surfaces are commonly used in the design of molded optics and the systems that employ them. In the next sections we review two such surfaces, aspheres and diffractives.

1.6.1 Aspheric Surfaces

Aspheric surfaces, in keeping with their name, are surfaces that are not spherical. While any surface that is not spherical could technically be called aspheric, when we use this term we are usually referring to a specific type of surface, whose form is described by

$$z = \frac{ch^2}{1 + \sqrt{1 - (1+k)c^2h^2}} + Ah^4 + Bh^6 + Ch^8 + Dh^{10} + \ldots \tag{1.6}$$

where z is the sag of the surface, c is the base curvature (the reciprocal of the radius) at the vertex of the surface, h is the radial distance from the vertex, k is the conic constant of the surface, and A, B, C, D, etc., are the fourth-order, sixth-order, eighth-order, tenth-order, and so on, aspheric coefficients. The sag is the axial depth of the surface, at any given point, relative to a plane perpendicular to the axis at the vertex of the surface. This mathematical description of an aspheric surface is widely used and is a standard surface form in optical design software. This description consists of the equation for a spherical surface (when $k = 0$), with the addition of even-order terms.

If the aspheric coefficients (A, B, C, etc.) are all zero, but the conic constant, k, is nonzero, the surface is referred to as a conic surface. The value of the conic constant determines the type of conic surface. A conic surface with a conic constant of –1 is a paraboloid. Conic surfaces with conic values less than –1 are hyperboloids, while surfaces with conic constants between –1 and 0 are ellipsoids. Conic surfaces with positive conic constants (greater than zero) are oblate ellipsoids, meaning that the line joining their foci is perpendicular to the optical axis. Conic surfaces have the interesting property of providing perfect (aberration-free) imaging for a single point on the line connecting their foci. In the case of a paraboloid, the mirror images a point on axis at infinite object distance to a perfect point image at the focus of the paraboloid, while an ellipsoid provides perfect imagery between its two foci.

Conic surfaces have often been used in reflective astronomical telescopes. Their point-to-point imaging property allows the mirrors to be accurately measured during the fabrication process. When using conics within a multi-element system, we generally are not concerned just with their point-to-point imaging properties, but with their aberration contribution to the entire field of view. As an example, a Ritchey-Chrétien telescope uses two hyperbolic mirrors in order to provide aberration correction of both spherical aberration and (third-order) coma.

An example of the ability of an aspheric surface to provide aberration correction is seen in Figure 1.15. The figure shows a convex-plano lens, similar to the one in Figure 1.5, with the exception that the front surface has been changed from a sphere to an asphere. While the spherical convex-plano lens exhibited spherical aberration, the spherical aberration of the aspheric

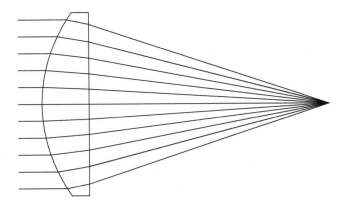

FIGURE 1.15
Aspheric convex-plano lens, showing correction of spherical aberration.

convex-plano lens has been corrected. This can be seen by the fact that all the rays, regardless of their radial height on the lens, come to focus at a common point along the axis. Although the spherical aberration has been corrected, other aberrations, such as coma, still exist. These aberrations would be evident by viewing the rays for an off-axis field. Thus, the use of an aspheric surface has improved the performance of the system, but not completely corrected all the aberrations.

In this example we used the aspheric surface to correct the spherical aberration of the element, as we were concerned only with the on-axis performance of the system. In general, most systems operate with some finite angular field, requiring control not only of spherical aberration, but coma, astigmatism, Petzval curvature, distortion, and for wideband or multiwavelength systems, chromatic aberrations as well.

The effect that an asphere has on the aberration contribution of a surface depends on its location within the system. If the aspheric surface is located at the aperture stop of the system or at the images of the aperture stop (the pupils), the aspheric surface will only directly affect the spherical aberration created at the surface. If the aspheric surface is located away from the aperture stop and pupils, it will directly affect the spherical aberration, coma, astigmatism, and distortion created at the surface. Note that we did not state it would directly affect the Petzval curvature, axial color, or lateral color, as these depend upon the base radius and material of the surface and not its asphericity. We use the term *directly* on purpose, in order to emphasize that the aspheric surface, acting independently, does not affect other surfaces. However, the asphere can indirectly change the aberration contribution of all the surfaces within the system. This is because providing specific aberration control at one surface allows the aberration contributions of the other surfaces to be adjusted, in order to provide the best overall system performance. For instance, if we place the aspheric surface at the aperture stop, allowing it to control spherical aberration, we do not need to be as concerned about the spherical aberration contribution of the other surfaces within the system. Instead, we can allow them to produce spherical aberration, which will be corrected by the stop-positioned aspheric surface, in order to correct or reduce other aberrations such as coma. The use and optimization of the aberration contributions of multiple surfaces, in order to balance and correct the aberrations of the total system, is the fundamental method used by the optical designer to obtain suitable optical performance.

The optimum placement of an aspheric surface within a system can be determined in a number of ways. Brute force, by iteratively placing an asphere on each of the surfaces of the system and running the optimization algorithm, is one method. A simple method of determining the best aspheric location is to examine which surface has the maximum beam extent. The surface with the largest beam extent is generally the position where the asphere will have the maximum leverage on aberration control. A more convenient method may be to use the features of the optical design code, which will show the designer the best location for an asphere. The popular optical design codes now each have a feature to perform this function.

When designing optical systems employing aspheric surfaces, as systems with molded optics often do, it is important that appropriate field coverage is applied. By applying appropriate field coverage, we mean that enough field angles should be entered into the optical design software to ensure that unexpected performance is not allowed to occur between adjacent field points. Because the optical design software optimizes the performance only at the defined field points, it is possible, if an inadequate number of field angles are entered, to have excellent performance at the designated fields at the expense of the performance between them. Figure 1.16 shows examples of a lens with too few fields defined (above) and a more appropriate number of fields defined (below) for designing with multiple aspheric surfaces. Once a design has been created, the performance should be evaluated at a number of field angles in addition to those that were originally defined and optimized, to ensure that there is no significant performance drop between them.

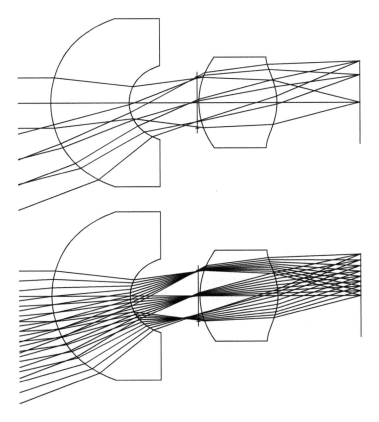

FIGURE 1.16
Increased numbers of fields for use with aspheric surfaces.

The pupil ray density used in the optimization of aspheric surfaces should also be increased from the default value. Similar to the situation of having too few fields, having too few rays can give a false impression of performance. The density of rays in the grid used to evaluate the merit function value can be user defined. During the design process and when evaluating the final design, several different (increasing) ray densities should be used to evaluate the merit function. If there are significant changes with increased ray density, reoptimization should occur with higher-density ray grids.

Additionally, when evaluating the design, the amount of asphericity of each aspheric surface should be evaluated. That is, the departure of the asphere from a best-fit spherical surface should be calculated. In some cases, particularly when there are multiple aspheric surfaces within the system, it may be found that one or more aspheric surfaces barely depart from a best-fit sphere. This is a logical result of the optimization algorithm within the optical design code. The optimization program will attempt to maximize the performance of the system, without regard to the amount of asphericity of a surface, unless specifically instructed not to do so. When a design has surfaces with limited departure from asphericity, the surfaces should be changed to spheres and the optimization algorithm rerun (after saving the initial design configuration file). Comparing the performance with and without the slightly aspheric surfaces will determine whether or not the added complexity of having such surfaces within the system is justified. While it is generally not more difficult to create a molded optic with an aspheric surface than with a spherical surface, there is no need to add unnecessary complexity into a system. It may even be appropriate, if aspheric surfaces are not required on a particular molded element, to replace it with a conventionally fabricated one. Evaluation of aspheric surfaces should also include determination of their manufacturability and producibility. Just because a surface can be designed on the computer does not mean that it can be readily produced or tested. Extremely deep or steeply sloped surfaces should be discussed with potential molders before the design is considered finished.

In addition to the even-order polynomial aspheric equation shown earlier, there are a number of other aspheric surface descriptions that can be used. These include odd aspheres, Zernike surface representations, and cubic splines. More recently, an alternate description of an aspheric surface, known as a Forbes asphere, has been gaining popularity. The Forbes description is meant to provide an aspheric representation that will create better optimized and more producible aspheric surfaces.[10] Whichever aspheric representation is used, discussion between the molder and designer should take place to ensure the surface definition is understood and the proper surface form is produced.

1.6.2 Diffractive Surfaces

While aspheric surfaces are used to provide correction of spherical aberration, coma, astigmatism, and distortion, diffractive surfaces are most often

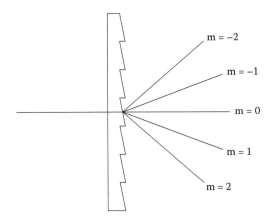

FIGURE 1.17
Schematic representation of a diffraction grating.

used to control the chromatic aberrations (axial and lateral color). Diffractive surfaces for molded optics, also known as diffractive optical elements (DOEs) or kinoforms, are typically composed of a microstructured pattern on the optical surface. An example of a diffractive surface that most readers have experienced is the diffraction grating, which is shown schematically in Figure 1.17. A diffraction grating usually consists of a series of equally spaced linear grooves. The grooves result in a multiwavelength beam incident on the grating being dispersed into its constituent colors. That is, the different wavelengths (colors) of light composing the beam leave the diffraction grating at varying angles. The grooves often have a triangular shape, which helps direct as much of each wavelength as possible into a single order (direction).

Diffractive surfaces are appropriately named, as they rely upon the principle of diffraction, which is associated with the wave nature of light. Much as multiple waves in a body of water can combine to form a resultant wave that is larger or smaller than each of the individual waves, multiple light waves can also combine to form a resultant wave. In the case of a molded optic diffractive surface, the waves that are combined are the segments of light that pass through each microstructure on the surface. By appropriately designing the microstructures, we can create a desired resultant wavefront that propagates after the surface.

Even though they are composed of microstructures and rely upon the wave nature of light, it is still possible to design and evaluate diffractive surfaces using ray tracing. However, instead of Snell's law, another equation is used to determine the direction of rays after passing through the surface. This equation is known as the grating equation:

$$d(\sin \theta_m - \sin \theta_i) = m\lambda \tag{1.7}$$

where d is the local period of the grating, θ_i is the angle of incidence, θ_m is the angle of diffraction for a given order, λ is the wavelength of light, and m is an integer denoting the grating order. Considering the example of the first-order ($m = 1$) diffracted light when the input rays are normally incident on the surface ($\theta_i = 0$), we can rearrange the grating equation to the form

$$\sin \theta_1 = \frac{\lambda}{d} \tag{1.8}$$

We can see that the sine of the angle of the diffracted ray is directly related to the wavelength of the ray. This is in contrast to a refractive surface, where Snell's law relates the sine of the angle not to the wavelength, but to the refractive index. Thus, the diffractive surface has a much larger variation in output ray angle with wavelength than a refractive surface, since the ratio of change in refractive index to change in wavelength is much less than 1. This is equivalent to a diffractive surface being much more dispersive than the material alone. An effective diffractive Abbe number, similar to that used to describe the dispersion of optical materials, can be calculated for the diffractive surface.[11] The effective Abbe number for a diffractive surface is defined as

$$V_{diff} = \frac{\lambda_d}{\lambda_F - \lambda_C} \tag{1.9}$$

where λ_d, λ_F, and λ_C are the wavelength values associated with n_d, n_F, and n_C.

Plugging in these wavelength values, we find that the effective visible Abbe number for a diffractive surface is –3.45. The magnitude of the diffractive Abbe number is much less than that of any material on the glass map shown in Figure 1.1. Since the sine of the diffracted angle is related directly to the wavelength, the grating bends (diffracts) red rays more than blue rays, which is the opposite of a refractive surface. This reversal of which color is bent more is the meaning of the negative sign in the Abbe number of the diffractive surface.

We discussed earlier that we could correct axial chromatic aberration by combining two elements in a doublet. We can perform a similar color correction by combining a refractive lens with a diffractive surface, often called a hybrid lens or a refractive-diffractive doublet. Because the Abbe number of the diffractive surface is negative, combining a positive power lens and a diffractive surface results in the diffractive having positive power. Besides completely correcting chromatic aberration, a defined amount of chromatic aberration can be designed into a hybrid lens (or conventional doublet). This is equivalent to creating a material with a desired dispersion. This can be quite useful when designing with molded plastic optics, as there are relatively few materials, and thus few Abbe numbers, to choose from.

A diffraction grating is usually made of parallel grooves and bends (not focuses) an incident beam. For a standard imaging system we instead want the diffractive surface to be rotationally symmetric and to focus, not just bend, the incident beam. Thus, most diffractives in imaging systems consist of annular, instead of linear, grooves. In order to bring the rays to a focus, as in a conventional positive lens, we need the rays at the outer portion of the surface to bend more than the rays at the inner portion. We see from the grating equation that the amount of angular change of a ray is related to the local grating period, with smaller periods causing greater angular changes. Thus, we can vary the grating period of the diffractive in order to create the desired amount of angular change as a function of radial position. The desire to bring rays to focus results in the grooves getting progressively smaller in width as the radial distance from the axis increases. This style of diffractive surface, with annular grooves decreasing in period as a function of radial distance, is the type most commonly seen in imaging systems employing them.

While a diffractive surface can provide significant improvement in system performance by correcting chromatic aberration, using a diffractive surface also brings with it associated costs. One of these costs is a potential transmission decrease, resulting from imperfect forming of the triangular profiles of the grooves. Rounded areas on the tops and bottoms of the grooves (see Figure 3.23), known as dead zones, can reduce transmission by acting as scattering sites. The impact of these dead zones can be determined by summing their annular areas. The ratio of total dead zone area to optical surface area is the transmission loss. For diffractive surfaces with a large number of rings, the transmission loss due to dead zones, if not controlled, can be significant. Limits or capabilities in regard to dead zone sizes should be discussed with molding vendors during the design process.

Another potential downside to using a diffractive surface is stray light due to imperfect diffraction efficiency. Diffraction efficiency is defined as the percentage of light incident on the diffractive microstructures that goes into the desired (design) order, usually the first order ($m = 1$). The heights of the microstructures (e.g., the triangular grooves) are normally determined according to

$$h = \frac{\lambda_0}{n_0 - 1} \tag{1.10}$$

where h is the height of the feature, λ_0 is the design wavelength, and n_0 is the refractive index of the material at the design wavelength. This equation assumes we are operating in the first order and that the diffractive surface is in air ($n_{air} = 1$). Because the microstructure has fixed height, it is optimal for only one wavelength (λ_0). As we move away from this wavelength, the diffraction efficiency drops, sending light into other diffractive orders. An equation for the diffraction efficiency and an associated plot is shown later, in Chapter 3. The light that goes into the nondesign orders can propagate through the system and end up on the image plane as stray light. Contrast reduction in the image can occur if the light from the nondesign orders is sufficiently spread out such that it creates an increased background level on the area of interest of the image plane. Reduced contrast generally results in less ability to see the fine details in the image. The amount of detail produced by an optical system is generally described by the modulation transfer function of the system. The reduction in MTF due to imperfect diffraction efficiency can be predicted, as has been described by multiple authors.[12,13]

When creating designs using DOEs, it is important to consider the impact of potential stray light features on the end user. It may be that such features result in unacceptable performance, forcing the removal of the diffractive surface and requiring alternate means of color correction or living with larger amounts of chromatic aberration. We strongly suggest prototyping and testing systems with diffractive surfaces to verify that their diffractive artifacts are acceptable. The prototype should be tested under similar conditions to which the system is expected to be used. Many times, under conditions of relatively uniform scene brightness, the artifacts will not be noticeable and the reduction in MTF allowable. However, under conditions with significant variation in brightness, such as an illuminated street lamp against a night sky, or high-contrast text (black letters and white background or the reverse), the artifacts may be found to be unacceptable.

Similar to an aspheric surface, the location of the diffractive surface within the system determines which (chromatic) aberrations it controls. A diffractive surface that is placed at the aperture stop (or equivalently at the pupils) introduces only axial color. A diffractive surface that is placed away from the aperture stop (or pupils) will affect both axial and lateral color. The further the diffractive is from the stop, the larger the leverage it has on lateral color.

As with aspheric surfaces, diffractive surfaces in a design should be evaluated for manufacturability and necessity. In regard to manufacturability, this generally means ensuring that the period of the grating, also known as the zone width or groove width, is sufficiently large. For production of diffractive mold masters by standard machining methods (e.g., diamond turning), zone widths of about 20 microns or larger are recommended. While smaller zones can be fabricated, these may require alternate methods, such as lithography, which may limit the sag of the surface the diffractive is placed upon, possibly forcing it to be planar. In addition, if the diffractive grooves become only several times the wavelength of light they will be used with, electromagnetic (vector) effects may come into play, invalidating the use of ray tracing in the design of the diffractive. With regard to necessity, it should be verified that adequate color correction cannot be achieved (for a given number of lenses) without the use of the diffractive surface. In general, it is best to use only one diffractive surface within a system. The use of more than one kinoform should be evaluated carefully, as multiple diffractives increase the potential for transmission loss and diffraction efficiency issues.

1.7 Tolerancing and Performance Prediction

No molded or conventionally made optical element will be perfect. There is always some variation between the part that is produced and the nominal design parameter values. Ideally, the system optical design would be highly insensitive to variation of the parts and assembly from the nominal condition. In reality, the variation of the part and assembly parameters usually limits the ultimate performance of the system. In order to develop systems that can realistically and cost-effectively be produced, a tolerance analysis needs to be performed. Tolerance analyses are used to evaluate the sensitivity of the system performance to the various tolerances associated with producing and assembling its elements. Additionally, they can provide predictions of the range of expected performance within a batch of manufactured systems.

1.7.1 Tolerance Sensitivity Analysis

Tolerance sensitivity analyses determine the relationship between the performance of the system and the individual tolerances on each parameter of the components and assembly. Examples of component parameters are surface radii and center thickness, while examples of assembly parameters are lens decentration and lens spacing. Tolerance sensitivity analysis can be performed manually, by changing individual parameters and determining the effect on system performance. However, this is generally not necessary, as optical design codes have tolerance analysis features that allow automated tolerance sensitivity evaluation. The designer can enter a set of expected tolerances and the design software will rapidly evaluate the performance against them. There are multiple default performance criteria that the software can evaluate against, including MTF and wavefront error. Other performance criteria, such as ensquared energy, are not usually default selections. These alternate types of criteria may need to be correlated to the default criteria or can be evaluated directly using user-defined software scripts. In addition to evaluating the performance change due to a set of defined tolerance values, the programs can also determine the magnitude of a tolerance that induces a certain drop in performance. Using the program in this manner allows the designer to determine which parameters are the most sensitive. This information can be fed back into the design process in an attempt to reduce or redistribute tolerance sensitivities, which generally produces more cost-effective and manufacturable designs.

1.7.2 Monte Carlo Analysis

A Monte Carlo tolerance analysis predicts the performance of as-built systems. Instead of evaluating the effect of each tolerance individually, the Monte Carlo evaluation "builds" representative production systems by randomly sampling from within the various tolerance ranges and distributions, then applying the sampled tolerances to the components and assembly. With the sampled imperfection values applied, the performance of the system is evaluated. Repeating this process many times leads to a distribution of predicted system performance. Comparing the predicted performance distribution to the pass/fail criteria allows expected yield values to be obtained. Monte Carlo evaluations generally provide more reliable performance distributions than the predictions from a tolerance sensitivity analysis. This is because more realistic systems are evaluated, as opposed to the statistical evaluation performed during the tolerance sensitivity analysis. However, Monte Carlo evaluations do take significantly longer to perform. We recommend using both tolerance sensitivity and Monte Carlo evaluations during the development of the design. The sensitivity analyses can provide feedback into which tolerances need to be controlled, while the Monte Carlo analysis can estimate performance range and yield.

1.7.3 Image Simulation

As seen above, modern optical design and analysis programs have features that also allow the prediction of system performance through image simulation. The image simulation features take an input image, such as a bitmap, and "pass it through" the optical system by applying ray tracing, diffraction calculations, and convolution operations. By processing images through the nominal design, as well as through representative as-built systems (which can be obtained through the Monte Carlo process above), the customer and designer can be provided with visual examples of the expected range of performance of the manufactured systems. Most customers (and most designers as well) have difficulty relating an image quality metric such as MTF or wavefront error to the actual system imaging performance. In this respect, the use of image simulation features allows a more intuitive feeling for the system performance. The input image can be provided by the customer or selected based on the intended system application. For instance, a cell phone camera may be used to take pictures of a building when touring a university. As with the simulations seen earlier, the input image can be processed through competing designs, allowing direct cost/performance comparisons. Based on this, we believe that image simulation features are excellent tools for use in design trades for systems with (and without) molded optics.

References

1. Smith, W. J. 2008. *Modern optical engineering*. New York: McGraw-Hill.
2. Hecht, E. 2001. *Optics*. Reading, MA: Addison-Wesley.

3. Fischer, R. E., B. Tadic-Galeb, and P. R. Yoder. 2008. *Optical system design*. New York: McGraw-Hill.
4. Jenkins, F., and H. White. 2001. *Fundamentals of optics*. New York: McGraw-Hill.
5. Shannon, R. R. 1997. *The art and science of optical design*. New York: Cambridge University Press.
6. Smith, W. J. 2004. *Modern lens design*. New York: McGraw-Hill.
7. Kingslake, R., and R. B. Johnson. 2009. *Lens design fundamentals*. San Diego: Academic Press.
8. Kidger, M. J. 2002. *Fundamental optical design*. Bellingham, WA: SPIE Press.
9. Kidger, M. J. 2004. *Intermediate optical design*. Bellingham, WA: SPIE Press.
10. Forbes, G. W. 2007. Shape specification for axially symmetric optical surfaces. *Optics Express* 15:5218–26.
11. Stone, T., and N. George. 1988. Hybrid diffractive-refractive lenses and achromats. *Applied Optics* 32:2295–302.
12. Londoño, C., and P. P. Clark. 1992. Modeling diffraction efficiency effects when designing hybrid diffractive lens systems. *Applied Optics* 31:2248–52.
13. Buralli, D. A., and G. M. Morris. 1992. Effects of diffraction efficiency on the modulation transfer function of diffractive lenses. *Applied Optics* 31:4389–96.

FIGURE 1.11
Input (object) for image simulation. (Courtesy of Dr. Eric Fest.)

FIGURE 1.12
Simulated image using a single-element system.

FIGURE 1.13
Simulated image using a three-element (triplet) system.

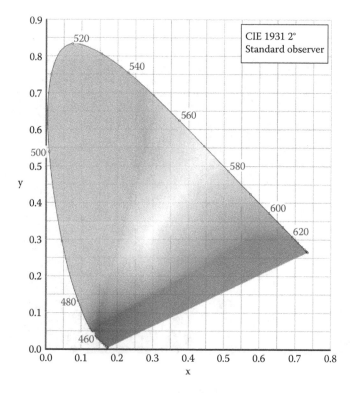

FIGURE 2.5
The chromaticity diagram for the CIE 1931 2° coloring-matching functions.

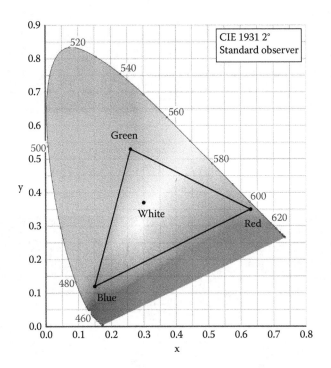

FIGURE 2.8
The gamut and white point of the microdisplay illustrated on the CIE 1931 chromaticity diagram.

FIGURE 4.15
Simulated image for the nominal cell phone camera lens design.

FIGURE 4.16
Simulated image for the cell phone camera lens with 10-micron surface decentrations applied.

2

Visual Optics

Jim Schwiegerling

CONTENTS

2.1 Introduction

The design of optical systems requires an understanding of the conditions and environment under which the system will be used. In addition to the performance of the optics, the characteristics of the scene and its illumination, as well as the capabilities of the final image sensor, need to be taken into consideration. Failure to incorporate the scene and sensor into the design process can lead to mismatches between components and unnecessary under- or overspecification of component performance. As a simple example of this mismatch, consider a diffraction-limited F/2 lens coupled with a 1/3-inch Video Graphics Array (VGA) Charged-Coupled Device (CCD) array as the image sensor. A diffraction-limited system has an Airy pattern as its point spread function. The size of this spot is proportional to the first zero crossing of the Airy disk. For a wavelength of 0.5896 µm, the diffraction-limited spot diameter produced by the lens is given by

$$\text{Spot diameter} = 2.44\lambda(F/\#) = 2.8 \ \mu\text{m} \qquad (2.1)$$

The image sensor, however, has overall rectangular dimensions of 4.8 × 3.6 mm and 640 × 480 pixels. Each pixel is therefore 7.5 µm square. Consequently, there is a performance mismatch between the optics and the sensor, where the optics can resolve features approximately three times smaller than what the sensor can resolve. Such a mismatch suggests that either the performance of the optics can be reduced or the pixel density of the sensor can be increased, depending on the specific imaging application.

Optical systems incorporating molded optical elements often have the eye as the final image sensor. Examples of these systems include simple magnifiers, telescopes, microscopes, and head-mounted displays. In this chapter, the ramifications of using the eye as the system detector are explored and the optical limitations of the eye and the human visual system as a whole will be illustrated. Here, the spectral sensitivity of the photoreceptors, the imaging capabilities of the eye under different lighting conditions, and the limitations chromatic and monochromatic aberrations impose on visual performance will be explored in detail. Furthermore, a brief introduction to colorimetry is provided. The goal of the chapter is to provide insight into the eye's optical performance so that external optical systems can be appropriately designed to couple with the eye. To illustrate the concepts outlined in this chapter, the design of a simple head-mounted display system is included.

2.2 The Human Eye and Visual System

The human eye is a globe-shaped system roughly 25 mm in diameter. It consists of two optical elements, the cornea and the gradient index crystalline lens. The optical elements form the boundaries of two distinct chambers: the anterior chamber and the posterior chamber. The anterior chamber is bounded on the front by the cornea, which is the transparent membrane that is visible on the outside of the eye. The cornea provides roughly two-thirds of the optical power of the eye. The anterior chamber is filled with a water-like substance called the aqueous humor that circulates and provides nutrients to the cornea. Also in the anterior chamber is the iris, which is the colored diaphragm visible through the cornea. The iris has the ability to dilate and contract the size of its opening from roughly 2 mm to 8 mm in response to different lighting conditions. The iris serves as the aperture stop of the eye. The crystalline lens, immediately behind the iris, isolates the anterior chamber from the posterior chamber. The crystalline lens has an onion-like structure in which multiple shells surround its center, creating a lens with a gradient refractive index profile in both the axial and radial directions. In addition, the crystalline lens, at least at younger ages, is flexible and can change its shape and power in response to contraction of the ciliary muscle that surrounds its equator. The space between the crystalline lens and the back of the eyeball is known as the posterior chamber. It is filled with a jelly-like substance known as vitreous humor. At the back of the eyeball, the interior of the globe serves as the image plane of the eye. The interior of the globe is lined with the retina, which is a layer of photosensitive cells

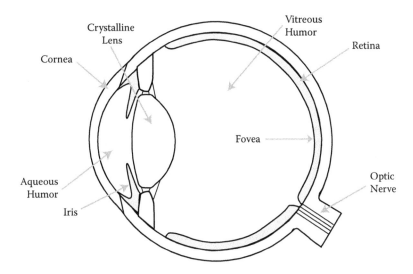

FIGURE 2.1
Top view of a cross section of a human eye.

known as photoreceptors. The retina further consists of additional cells and wiring required to relay the light patterns detected by the photoreceptors to the brain for further processing. Figure 2.1 illustrates the basic elements of the eye.

As with other features of the human form, there is a large variation in the size and shape of the various components of the eye. To adequately design optical systems that couple to the eye, it is useful to have a model of the eye that represents its average dimensions and performance. In this manner, the optical system can be designed to cover a large array of human observers, while only failing to account for people in the tails of the population distribution. Schematic eye models have evolved over time from simple spherical models matching the basic cardinal points of the average eye to aspheric multielement models that incorporate monochromatic and chromatic aberrations. In addition, it is possible to customize eye models by measuring surface shapes and ocular aberrations from an individual eye and creating a corresponding schematic eye model. However, this technique is too tailored for application to a widely deployed optical system. Here, a step back from this customized schematic eye will be taken to ensure applicability to a broad range of people. The eye model must have average surface curvatures, conic constants, and separations found in the population. The conic constants allow adjustments to give a population average value of ocular spherical aberration. The material indices must have a realistic dispersion to match ocular chromatic aberration. By adding these features, realistic and broadly applicable models can be created. In general, the optical surfaces of the eye are nonrotationally symmetric and noncoaxial. Furthermore, the crystalline lens has a gradient refractive index profile. While these features can be

TABLE 2.1

Arizona Eye Model

Surface	Radius (mm)	Conic Constant	Index n_d	Abbe Number	Thickness (mm)
Anterior cornea	7.800	−0.250	1.377	57.1	0.550
Posterior cornea	6.500	−0.250	1.337	61.3	2.970
Anterior crystalline lens	12.000	−7.519	1.42	51.9	3.767
Posterior crystalline lens	−5.225	−1.354	1.336	61.1	16.713
Retina	−13.400	0.000			

incorporated into a schematic eye model, the additional complexity of these models is not warranted in a general optical system design. Consequently, the eye's optical surfaces will be assumed to be rotationally symmetric about a single optical axis, and the refractive index of the crystalline lens will be assumed to be a homogeneous "effective" index of refraction. There are several schematic eye models in the literature that meet these criteria.[1-4] For the purposes of this chapter, the Arizona Eye model[5] will be used. The parameters of this schematic eye model with the eye focused at infinity are given in Table 2.1. The values from this table are easily entered into modern ray-tracing packages for optical analysis of the performance of the eye. A first-order analysis of the eye model provides information regarding the cardinal points and pupils of the system. These values are summarized in Table 2.2, where all of the distances are relative to the vertex V of the anterior corneal surface. Note that the nodal points do not coincide with their respective principal points, and that the anterior and posterior focal lengths are different. These effects are due to the refractive index of the image space being that of vitreous humor, while the index of the object space is unity. As can be seen from Table 2.2, the effective focal length, f_{eye}, of the eye is

$$f_{eye} = VP - VF = \frac{VF' - VP'}{1.336} = 16.498 \text{ mm} \qquad (2.2)$$

TABLE 2.2

Cardinal Points and Pupil Positions of Arizona Eye Model

Line Segment	Distance (mm)	Line Segment	Distance (mm)
VF	−14.850	VN	7.191
VF'	24.000	VN'	7.502
VP	1.648	VE	2.962
VP'	1.960	VE'	3.585

Note: V = vertex of anterior corneal surface; F, F' = front/rear focal points; P, P' = front/rear principal points; N, N' = front/rear nodal points; E, E' = entrance/exit pupil positions.

While the surfaces of the eye model are assumed rotationally symmetric and coaxial, the entire model can rotate relative to an external optical system. The eye has the ability to rotate within its socket, allowing the gaze angle to fixate on different objects in the scene. The center of rotation of the eye is approximately 7 mm behind the corneal vertex, roughly coinciding with the front nodal point. The design example at the end of the chapter examines a head-mounted display system. Looking into the system, the eye can rotate about this center of rotation to look at different locations on the display.

A distinction needs to be made between the eye and the human visual system. The eye is the imaging system of human vision. It consists of optical elements that relay light onto an image plane. The eye suffers from aberrations, diffraction, dispersion, and transmission losses in the same manner as any other optical system. The human visual system, on the other hand, includes all of the components that allow us to see. These components include the imaging system of the eye, the photoreceptors that record the light incident on the image plane and convert it to an electrical signal, and the neural processes, which in turn interpret these signals and convert them into the perceived image. These additional components of the visual system can be classified. There are two major types of photoreceptors, the rods and the cones. Furthermore, the cones themselves consist of three distinct classes, L-cones, M-cones, and S-cones. Each type and class of photoreceptor has a specific spectral response and a specific range of illumination levels over which they perform well. The rods are sensitive to a band of wavelengths toward the blue end of the visible spectrum. They are also only active in dim lighting conditions (roughly moonlight and lower illumination levels). These lighting conditions are known as scotopic conditions. The cones, on the other hand, are responsible for vision at higher illumination levels, as well as for providing color vision. These lighting conditions where the cones are active are known as photopic conditions. The L-cones are sensitive to a band of wavelengths in the red end of the spectrum, the M-cones are sensitive to a wavelength band in the green portion of the spectrum, and the S-cones are sensitive to the blue end of the visible spectrum. The L, M, and S designations refer to the cone type's spectral sensitivity as long wavelength, middle wavelength, and short wavelength, respectively.

The final components of the human visual system are the neural processes. Each type of photoreceptor has the ability to absorb a photon and trigger an electrochemical signal that is relayed to the brain. The probability of absorbing the photon is related to the spectral sensitivity of the photoreceptor and the ambient lighting conditions. Once a photon is absorbed, the subsequent signals cannot distinguish the wavelength that started the reaction. The perception of color is only achieved by comparing the relative signal response from different photoreceptor types. The absorption of the photon initiates a signal and further neural processing that ultimately leads to perception. The neural processes start at the level of the retina where cells connect the photoreceptors to neural cells located in the brain in areas such as the lateral geniculate nucleus and the visual cortex. These neural processes are responsible for displaying the image in our mind's eye, as well as processing and enhancing the image that is perceived. Contrast enhancement, edge detection, and gain adjustment are all regulated by these processes. Furthermore, the perceived image is analyzed to provide feedback mechanisms that tell the head to rotate to acquire a target, that rotate the eye in its socket to align with the target, that trigger the ciliary muscle to adjust the crystalline lens to focus on an object, and to signal the iris to adjust its aperture size in response to the ambient lighting conditions and the proximity of the target. The human visual system is a remarkable system. The eye has a nearly 180° field of view. In addition, the young eye can continuously vary its focus from infinitely far objects to objects only 40 mm from the eye. This latter focus adjustment is called accommodation and is achieved by changing the shape and the refractive index distribution of the crystalline lens. The visual system can respond to illuminations levels covering ten orders of magnitude aided only slightly by the size of the iris and more notably by the absolute sensitivities of the photoreceptors. The neural processes enhance contrast and detect edges to aid in the detection and isolation of objects from the background. The breadth of capabilities of the human eye and the visual system are well beyond the scope of this chapter. Consequently, several assumptions will be made in the ensuing discussion of designing molded optical systems that incorporate the eye as the final detector. These assumptions will hold for typical design situations.

The first assumption is that only photopic illumination levels will be considered. Typically, an optical system will present information to the eye that is sufficiently bright for comfortable interpretation of the image. For example, the display in a head-mounted display may operate at a peak luminance of 100 cd/m^2, which is well within the photopic range. The display is sufficiently bright to reveal features in the dim portions of the image while not so bright as to cause physical discomfort to the viewer. As a result of this assumption, the effects of the rods and scotopic vision will be ignored. For the second assumption, only foveal vision will be considered. The fovea is a small region in the central portion of the retina. In a conventional optical system, the fovea would correspond to the neighborhood around the intersection of the optical axis with the paraxial image plane. The fovea provides the maximum resolution due to small, tightly packed cones, each with its own connection to the downstream neural processing. Outside the fovea, the size of the photoreceptors progressively increases and the signals from multiple photoreceptors are combined prior to transmission to the brain. Consequently, the resolution capabilities of the retina rapidly decrease when moving away from the fovea. This effect is due to a reduction in the actual sampling density from the increased size of the photoreceptors, as well as a decrease in the effective sampling density from integrating the signal from multiple photoreceptors. The consequence of this second assumption is that the performance of the optical system for peripheral vision will be ignored.

Typically, poor performance of the optical system for larger fields is offset by reduced performance of the peripheral vision. One caveat of this assumption is that the eye is mobile. In other words, the eye can rotate to bring an off-axis object into alignment with the fovea. The performance of the optical system therefore must be sufficient to handle these eye rotations. The third assumption is that the eye can bring the image produced by the molded optical system into focus onto the retina. For a perfect eye, the light entering the eye should be collimated. However, people suffering from near- or farsightedness have an intrinsic defocus error in their eye. The external optical system needs to compensate for this effect. In the head-mounted display example, the display would typically be located at the front focal plane of the eyepiece such that the rays emerging from a point on the display are collimated as they exit the eyepiece. Adjusting the distance between the display and the eyepiece causes these rays to diverge or converge as they emerge from the eyepiece, which in turn can compensate for the defocus of an imperfect eye. Accommodation can also be used to aid in the focus adjustment, but only moderate levels of accommodation should be used, as extended durations and levels of accommodation cause eye strain and fatigue. Furthermore, older people lose the ability to accommodate, so this mechanism may not always be available. The final assumption is that the eye is healthy. There are a variety of ocular pathologies that can confound the performance of any external optical system. For example, cataracts are opacities in the crystalline lens that cause excessive amounts of scattered light within the eye. The scattered light creates a veiling illumination onto the retina, ultimately degrading visual performance. Furthermore, diseases such as glaucoma and macular degeneration cause damage to the photoreceptors and the wiring in the retina. Even if high-quality images can be delivered to the retina, inadequate response from the detection mechanisms degrades the perceived image. Finally, amblyopia is a problem in which the eye did not have sufficiently high-quality images at a young age to establish the neural connections required to process high spatial frequency information. Even if these errors are corrected later in life, the neural processing lags and cannot readily handle all the information provided by the eye, resulting in a degraded perceived image. The effects of such diseases and afflictions will be ignored in the ensuing analysis and only clear, smooth optical surfaces and properly functioning retinal and neural processes will be considered.

2.3 Photopic Response and Colorimetry

The Commission Internationale de l'Éclairage (CIE) is an international organization devoted to understanding and developing standards pertaining to light and lighting. A portion of their mission has been devoted to creating

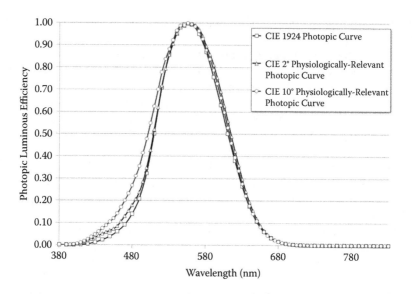

FIGURE 2.2

Photopic response curves. The physiologically relevant definition corrects problems with previous standards, especially at shorter wavelengths.

standards to describe the human perception of light and color. Among these standards are the 1924 2° photopic luminosity curve, denoted as V(λ) by convention.[6] This curve represents the spectral response of the eye under photopic conditions for small (2°) targets. While this standard is commonly presented in many optics texts, there are some problems with the curve, especially for the shorter wavelengths.[7] More recently, the CIE has corrected these short-wavelength problems with the adoption of physiologically relevant 2° and 10° luminous efficiency functions.[8,9] Figure 2.2 compares the various definitions of the photopic response curves and their corrections. These curves peak at a wavelength of approximately 555 nm and fall off for both shorter and longer wavelengths. For the design process, this shape suggests that the center of the visible spectrum should be weighted more heavily than the red or blue wavelength regions when optimizing the performance of an external optical system that will be coupled to the eye. This low response at the ends of the visible spectrum also suggests that sensitivity to longitudinal chromatic aberration is reduced when the eye is used as the final system sensor relative to typical focal plane arrays.

Colorimetry is the measurement or quantification of light as it pertains to the perception of human color vision.[10] The CIE has a long history of standardizing color-matching functions (CMFs), which form the basis for colorimetry. The CMFs are a description of how a typical observer with normal color vision would mix three primary lights to match a pure monochromatic light of a given wavelength. The CIE 1931 2° CMFs are based on experiments performed by Wright[11] and Guild.[12] In their experiments, a bipartite field is

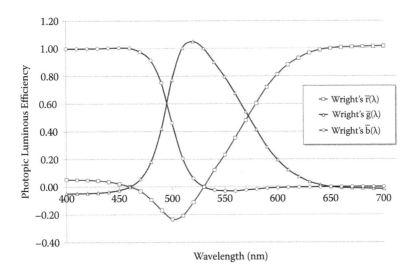

FIGURE 2.3
The $\bar{r}(\lambda)$, $\bar{g}(\lambda)$, and $\bar{b}(\lambda)$ color-matching functions from Wright (1928) and Guild (1931).

presented to a series of observers. A bipartite field is a split screen, where one half of the screen is illuminated with a monochromatic test light of a given wavelength and the other half of the screen is illuminated by a mixture of three primary light sources. For these experiments, monochromatic primary lights with wavelengths of 0.444, 0.526, and 0.645 μm were used. The observer is able to adjust the intensities of the primary lights until their mixture matches the test light. The relative intensities of the primaries are noted and the process is then repeated for test lights throughout the visible spectrum. The relative intensities of the primaries as a function of wavelength form the color-matching functions, typically denoted by $\bar{r}(\lambda)$, $\bar{g}(\lambda)$, and $\bar{b}(\lambda)$. Figure 2.3 shows the CMFs from these experiments. One curious feature of Figure 2.3 is that $\bar{r}(\lambda)$ is negative for a portion of the visible spectrum. When performing the color-matching experiments, there are some wavelengths in which combinations of the three primaries cannot match the monochromatic test light. However, if one of the primaries is removed from the primary side of the field and added to the test side of the field, then a match can be achieved. The negative portions of the CMF represent these cases of combining one of the primaries with the test light.

The primaries chosen in the Wright and Guild experiments are by no means unique. Any three independent light sources, either monochromatic or broadband, can be used as primaries. The independence requirement means that one of the light sources is not a linear combination of the remaining two primaries. Any perceivable color can be represented as a unique linear combination of the primary lights, with the caveat that sometimes one of the primaries needs to be combined with the test light to get a match. As a result of being able to represent perceived colors as a combination of three light sources, the concept of a three-dimensional color space can be used to represent colors. Each point or coordinate in this space is a specific combination of the weights of each primary light. Colors that can be perceived form a three-dimensional solid within this space, with perceptible colors inside or on the surface of the solid and imperceptible "colors" residing outside the solid. The coordinates of a point in this color space are determined by projecting the spectral power distribution $P(\lambda)$ of a given source onto each of the color-matching functions. For example, a point (R, G, B) in the color space defined by the Wright and Guild CMFs is given by

$$R = \int P(\lambda)\bar{r}(\lambda)d\lambda \quad G = \int P(\lambda)\bar{g}(\lambda)d\lambda \quad B = \int P(\lambda)\bar{b}(\lambda)d\lambda \quad (2.3)$$

This color space is sometimes called the CIE RGB color space. There is a wide assortment of RGB color spaces, and the term is sometimes used cavalierly. These various spaces have often been tied to specific devices and have different primaries depending on the application. When using RGB color spaces, care should be taken to understand which space is being used. In the ensuing analysis, only the CIE RGB color space as defined by the Wright and Guild CMFs will be considered. The linear nature of the CIE RGB color space means that a new color space with a new set of primaries can be created as a linear combination of the old primaries. Thus, an infinite number of color spaces can be created by transforming the original primaries into a new set of primaries. With knowledge of these properties, CIE sought to create a standardized color space by transforming the Wright and Guild results into a new set of color-matching functions. The most logical color space would be one in which the axes represented the response of the three cone types. However, at the time of this research in the early part of the twentieth century, the spectral sensitivities of the cones had not been isolated, and only recently have accurate measures of these data been obtained. Due to this limitation, the CIE chose from the infinite number of possibilities a transformation to a set of CMFs with two desirable properties. These CMFs are represented as $\bar{x}(\lambda)$, $\bar{y}(\lambda)$, and $\bar{z}(\lambda)$. The first property is that the CMFs are strictly positive, and the second property is that $\bar{y}(\lambda)$ is the photopic luminosity curve. In this manner, a test wavelength can always be matched without having to move one of the primaries to the test side of the bipartite field, and the "brightness" information is encoded along with the color information. The consequence of these choices, though, is that the primaries are imaginary in that they reside in a portion of the color space that lies outside of what we can perceive. However, suitable combinations of these imaginary primaries lead to points within the color solid. The CIE adopted a standard set of CMFs in 1931 that are suitable for small field (2°) situations.[13] A similar set of CMFs for large fields (10°) were adopted by the CIE in 1964, denoted as $\bar{x}_{10}(\lambda)$, $\bar{y}_{10}(\lambda)$, and $\bar{z}_{10}(\lambda)$.[14] Finally, there are several other sets of CMFs available that are aimed

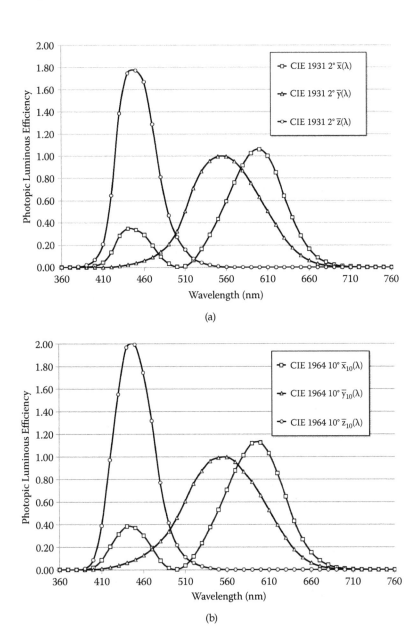

FIGURE 2.4
(a) The CIE $\bar{x}(\lambda)$, $\bar{y}(\lambda)$, and $\bar{z}(\lambda)$, and (b) $\bar{x}_{10}(\lambda)$, $\bar{y}_{10}(\lambda)$, and $\bar{z}_{10}(\lambda)$ color-matching functions.

at improving the CIE standard sets.[7,14–15] However, their adoption has been limited. Figure 2.4 shows the $\bar{x}(\lambda)$, $\bar{y}(\lambda)$, and $\bar{z}(\lambda)$, and $\bar{x}_{10}(\lambda)$, $\bar{y}_{10}(\lambda)$, and $\bar{z}_{10}(\lambda)$ CMFs, and their numerical values are summarized in Table 2.3. The coordinates of a point in the color space defined by the 1931 CIE 2° CMFs or the 1964 CIE 10° CMFs are given in similar fashion to Equation 2.3, namely:

TABLE 2.3

Table of the CIE Color-Matching Functions for the 2° and 10° Cases

λ (nm)	$\bar{x}(\lambda)$	$\bar{y}(\lambda)$	$\bar{z}(\lambda)$	$\bar{x}_{10}(\lambda)$	$\bar{y}_{10}(\lambda)$	$\bar{z}_{10}(\lambda)$
380	0.00137	0.00004	0.00645	0.00016	0.00002	0.00070
385	0.00224	0.00006	0.01055	0.00066	0.00007	0.00293
390	0.00424	0.00012	0.02005	0.00236	0.00025	0.01048
395	0.00765	0.00022	0.03621	0.00724	0.00077	0.03234
400	0.01431	0.00040	0.06785	0.01911	0.00200	0.08601
405	0.02319	0.00064	0.11020	0.04340	0.00451	0.19712
410	0.04351	0.00121	0.20740	0.08474	0.00876	0.38937
415	0.07763	0.00218	0.37130	0.14064	0.01446	0.65676
420	0.13438	0.00400	0.64560	0.20449	0.02139	0.97254
425	0.21477	0.00730	1.03905	0.26474	0.02950	1.28250
430	0.28390	0.01160	1.38560	0.31468	0.03868	1.55348
435	0.32850	0.01684	1.62296	0.35772	0.04960	1.79850
440	0.34828	0.02300	1.74706	0.38373	0.06208	1.96728
445	0.34806	0.02980	1.78260	0.38673	0.07470	2.02730
450	0.33620	0.03800	1.77211	0.37070	0.08946	1.99480
455	0.31870	0.04800	1.74410	0.34296	0.10626	1.90070
460	0.29080	0.06000	1.66920	0.30227	0.12820	1.74537
465	0.25110	0.07390	1.52810	0.25409	0.15276	1.55490
470	0.19536	0.09098	1.28764	0.19562	0.18519	1.31756
475	0.14210	0.11260	1.04190	0.13235	0.21994	1.03020
480	0.09564	0.13902	0.81295	0.08051	0.25359	0.77213
485	0.05795	0.16930	0.61620	0.04107	0.29767	0.57060
490	0.03201	0.20802	0.46518	0.01617	0.33913	0.41525
495	0.01470	0.25860	0.35330	0.00513	0.39538	0.30236
500	0.00490	0.32300	0.27200	0.00382	0.46078	0.21850
505	0.00240	0.40730	0.21230	0.01544	0.53136	0.15925
510	0.00930	0.50300	0.15820	0.03747	0.60674	0.11204
515	0.02910	0.60820	0.11170	0.07136	0.68566	0.08225
520	0.06327	0.71000	0.07825	0.11775	0.76176	0.06071
525	0.10960	0.79320	0.05725	0.17295	0.82333	0.04305
530	0.16550	0.86200	0.04216	0.23649	0.87521	0.03045
535	0.22575	0.91485	0.02984	0.30421	0.92381	0.02058
540	0.29040	0.95400	0.02030	0.37677	0.96199	0.01368
545	0.35970	0.98030	0.01340	0.45158	0.98220	0.00792
550	0.43345	0.99495	0.00875	0.52983	0.99176	0.00399
555	0.51205	1.00000	0.00575	0.61605	0.99911	0.00109
560	0.59450	0.99500	0.00390	0.70522	0.99734	0.00000
565	0.67840	0.97860	0.00275	0.79383	0.98238	0.00000
570	0.76210	0.95200	0.00210	0.87866	0.95555	0.00000
575	0.84250	0.91540	0.00180	0.95116	0.91518	0.00000

TABLE 2.3 (continued)

Table of the CIE Color-Matching Functions for the 2° and 10° Cases

λ (nm)	$\bar{x}(\lambda)$	$\bar{y}(\lambda)$	$\bar{z}(\lambda)$	$\bar{x}_{10}(\lambda)$	$\bar{y}_{10}(\lambda)$	$\bar{z}_{10}(\lambda)$
585	0.97860	0.81630	0.00140	1.07430	0.82562	0.00000
590	1.02630	0.75700	0.00110	1.11852	0.77741	0.00000
595	1.05670	0.69490	0.00100	1.13430	0.72035	0.00000
600	1.06220	0.63100	0.00080	1.12399	0.65834	0.00000
605	1.04560	0.56680	0.00060	1.08910	0.59388	0.00000
610	1.00260	0.50300	0.00034	1.03048	0.52796	0.00000
615	0.93840	0.44120	0.00024	0.95074	0.46183	0.00000
620	0.85445	0.38100	0.00019	0.85630	0.39806	0.00000
625	0.75140	0.32100	0.00010	0.75493	0.33955	0.00000
630	0.64240	0.26500	0.00005	0.64747	0.28349	0.00000
635	0.54190	0.21700	0.00003	0.53511	0.22825	0.00000
640	0.44790	0.17500	0.00002	0.43157	0.17983	0.00000
645	0.36080	0.13820	0.00001	0.34369	0.14021	0.00000
650	0.28350	0.10700	0.00000	0.26833	0.10763	0.00000
655	0.21870	0.08160	0.00000	0.20430	0.08119	0.00000
660	0.16490	0.06100	0.00000	0.15257	0.06028	0.00000
665	0.12120	0.04458	0.00000	0.11221	0.04410	0.00000
670	0.08740	0.03200	0.00000	0.08126	0.03180	0.00000
675	0.06360	0.02320	0.00000	0.05793	0.02260	0.00000
680	0.04677	0.01700	0.00000	0.04085	0.01591	0.00000
685	0.03290	0.01192	0.00000	0.02862	0.01113	0.00000
690	0.02270	0.00821	0.00000	0.01994	0.00775	0.00000
695	0.01584	0.00572	0.00000	0.01384	0.00538	0.00000
700	0.01136	0.00410	0.00000	0.00958	0.00372	0.00000
705	0.00811	0.00293	0.00000	0.00661	0.00256	0.00000
710	0.00579	0.00209	0.00000	0.00455	0.00177	0.00000
715	0.00411	0.00148	0.00000	0.00314	0.00122	0.00000
720	0.00290	0.00105	0.00000	0.00217	0.00085	0.00000
725	0.00205	0.00074	0.00000	0.00151	0.00059	0.00000
730	0.00144	0.00052	0.00000	0.00104	0.00041	0.00000
735	0.00100	0.00036	0.00000	0.00073	0.00028	0.00000
740	0.00069	0.00025	0.00000	0.00051	0.00020	0.00000
745	0.00048	0.00017	0.00000	0.00036	0.00014	0.00000
750	0.00033	0.00012	0.00000	0.00025	0.00010	0.00000
755	0.00023	0.00008	0.00000	0.00018	0.00007	0.00000
760	0.00017	0.00006	0.00000	0.00013	0.00005	0.00000
765	0.00012	0.00004	0.00000	0.00009	0.00004	0.00000
770	0.00008	0.00003	0.00000	0.00006	0.00003	0.00000
775	0.00006	0.00002	0.00000	0.00005	0.00002	0.00000
780	0.00000	0.00000	0.00000	0.00003	0.00001	0.00000

$$X = \int P(\lambda)\bar{x}(\lambda)d\lambda \quad Y = \int P(\lambda)\bar{y}(\lambda)d\lambda \quad Z = \int P(\lambda)\bar{z}(\lambda)d\lambda \quad (2.4)$$

The 3-tuple (X, Y, Z) are referred to as the tristimulus value. Specific spectral power distributions $P(\lambda)$ or the entire color solid can be displayed by plotting the tristimulus values on a three-dimensional plot. The tristimulus coordinates of Equation 2.4 are related to the Wright and Guild CIE RGB coordinates of Equation 2.3 through a linear transformation. This relationship can be summarized by the matrix expression

$$\begin{pmatrix} X \\ Y \\ Z \end{pmatrix} = \begin{pmatrix} 0.4887 & 0.3107 & 0.2006 \\ 0.1762 & 0.8130 & 0.0108 \\ 0.0000 & 0.0102 & 0.9898 \end{pmatrix} \begin{pmatrix} R \\ G \\ B \end{pmatrix} \quad (2.5)$$

Of course, given the tristimulus values (X, Y, Z), the original CIE RGB coordinates can be recovered by inverting the matrix of Equation 2.5.

Spectrally pure colors (i.e., monochromatic) of wavelength λ_o have a spectral power distribution $P(\lambda) = \delta(\lambda - \lambda_o)$, where $\delta()$ is the Dirac delta function. Dirac delta functions have the following sifting property where the delta function samples or sifts out a specific value of a function:

$$f(x_o) = \int f(x)\delta(x - x_o)dx \quad (2.6)$$

Inserting the spectrally pure $P(\lambda) = \delta(\lambda - \lambda_o)$ into Equation 2.4 and using the sifting properties of delta functions shows that the tristimulus values of a spectrally pure color of wavelength λ_o are simply the CMFs sampled at the wavelength $(X, Y, Z) = (\bar{x}(\lambda_o), \bar{y}(\lambda_o), \bar{z}(\lambda_o))$. Conversely, a "white" spectral power distribution has the same value for all wavelengths, or $P(\lambda) = 1$. Equation 2.4 leads to $X = Y = Z = 1$ in this case, since the CIE CMFs are normalized to have unit area. An alternative means of displaying colorimetric data is projecting the three-dimensional tristimulus space onto a plane. The chromaticity coordinates (x, y, z) are defined as

$$x = \frac{X}{X+Y+Z} \quad y = \frac{Y}{X+Y+Z} \quad z = \frac{Z}{X+Y+Z} = 1 - x - y \quad (2.7)$$

Note that the chromaticity coordinate, z, is not independent of x and y. This dependency means that the colorimetric data can be represented with the two-dimensional plot of the chromaticity coordinates (x, y). The effect of the projection of the three-dimensional tristimulus values onto the two-dimensional chromaticity coordinates is a removal of luminance information from the representation. Consequently, the chromaticity coordinates

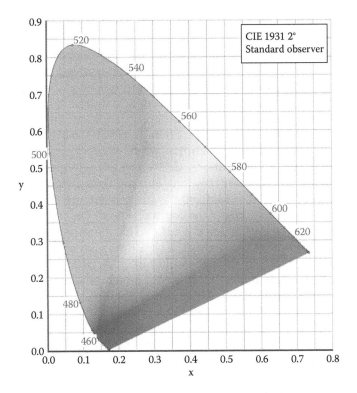

FIGURE 2.5 (See color insert.)
The chromaticity diagram for the CIE 1931 2° coloring-matching functions.

capture the hue and saturation of a given spectral power distribution, but lack information regarding its brightness. Figure 2.5 illustrates the chromaticity diagram for the CIE 1931 2° CMFs. In analyzing spectrally pure colors of wavelength λ_o, the tristimulus values were given by the delta-function-sampled CMFs. Incorporating this result into the chromaticity coordinates (x, y) of Equation 2.7 gives

$$x(\lambda_o) = \frac{\bar{x}(\lambda_o)}{\bar{x}(\lambda_o) + \bar{y}(\lambda_o) + \bar{z}(\lambda_o)} \qquad y(\lambda_o) = \frac{\bar{y}(\lambda_o)}{\bar{x}(\lambda_o) + \bar{y}(\lambda_o) + \bar{z}(\lambda_o)} \qquad (2.8)$$

The spectrally pure colors map out a horseshoe-shaped curve called the spectral locus, given by Equation 2.8. The opening of the horseshoe curve is bounded by a straight line connecting the chromaticity coordinates for violet wavelengths to red wavelengths. This line is called the alychne. All colors that can be perceived are contained within this bounded region of the chromaticity diagram. The white spectral power distribution with $P(\lambda) = 1$ corresponds to chromaticity coordinates $(x, y) = (0.333, 0.333)$. This point is called the equal energy white point, denoted as E. Colors perceived as neutral have chromaticity coordinates close to the white point, while saturated colors lead to chromaticity coordinates near the spectral locus. For a given hue, the line connecting the white point to the point on the spectral locus represents a continuous change in saturation, from unsaturated or neutral at the white point to fully saturated or pure at the spectral locus. The equal energy white point E is a theoretical construct. Real light sources do not have a perfectly constant spectral power distribution. In addition, the human visual system adapts to the ambient illumination. In effect, the human visual system performs color balancing such that white objects remain white under a variety of illumination conditions. Consequently, a unique white point is ill-defined in the chromaticity diagram, and multiple options for the white point exist. The CIE has standardized illuminants that are theoretical representations of real light sources. One common choice for a white point is an illuminant defined by the CIE known as D_{65}. This illuminant approximates a 6,500 K blackbody source or roughly the noontime solar spectrum. A second common choice defined by the CIE is illuminant A, which is representative of an incandescent light bulb. Illuminant D_{65} has chromaticity coordinates of (0.313, 0.329) for the CIE 1931 2° CMFs, while illuminant A has chromaticity coordinates (0.448, 0.407). Illuminant D_{65} resides near the equal energy white point, while illuminant A shifts the white point toward the red-orange portion of the spectral locus. In addition to standardizing the tristimulus color space, the CIE has also worked extensively in the area of detecting color differences. A fundamental problem of colorimetry is determining if two colors match. Stated another way, how close do two points, (X_1, Y_1, Z_1) and (X_2, Y_2, Z_2) in tristimulus space, need to be for the difference between the two colors to be imperceptible to a human observer? The color difference ΔE between these points can be defined in the typical Euclidean fashion as

$$\Delta E = \sqrt{(X_1 - X_2)^2 + (Y_1 - Y_2)^2 + (Z_1 - Z_2)^2} \qquad (2.9)$$

MacAdam[16] set up an experiment where an observer would view two colors of fixed luminance. One color was at a fixed coordinate in tristimulus space, while the other could be varied to approach the first color from different directions in the color space. With repeated testing, the mismatch between the two colors that the observer judges as indistinguishable defines the limitation of human perception to differentiate similar colors. MacAdam found two interesting results. First, the boundary enclosing indistinguishable colors tends to be an ellipse. This result means that the human ability to tell two colors apart depends on the direction the adjustable color approaches the target color. The second result is that the size of the ellipse depends on the absolute location within the tristimulus space. This latter result means, for example, that the just noticeable color difference between two green colors is different from the just noticeable color difference between two blue colors. In general, the just noticeable color differences for greens are larger than those for the reds, which in turn are larger than those for the blues in tristimulus space. These results suggest that the tristimulus space is warped such that equally distinguishable colors are nonuniformly spaced. Thus, the tristimulus space does not lend itself to easily predicting the likelihood of two colors matching based on their coordinates. The CIE has made several attempts to rectify spatial-dependent color differences, including the 1976 CIELUV and the 1976 CIELAB color space. These color spaces seek to provide a more perceptually uniform color space such that the distance between two points describes how different the colors are in luminance, chroma, and hue. Chroma is a measure of how different a color is from a gray of equivalent brightness. The 1976 CIELUV color space is one attempt at providing a perceptually uniform color space that has been standardized by the CIE. The CIELUV coordinates (L^*, u^*, v^*) can be calculated from the tristimulus values (X, Y, Z) or the chromaticity coordinates (x, y) with the following formulas. The subscript w denotes the values calculated with the tristimulus values and chromaticity coordinates of the white point.

$$L^* = 116\left(\frac{Y}{Y_w}\right)^{1/3} - 16 \quad \text{for} \quad \frac{Y}{Y_w} > 0.008856$$

$$L^* = 903.292\left(\frac{Y}{Y_w}\right) \quad \text{for} \quad \frac{Y}{Y_w} \le 0.008856$$

$$u^* = 13L^*\left[u' - u_w'\right]$$

$$v^* = 13L^*\left[v' - v_w'\right]$$

(2.10)

$$u' = \frac{4X}{X + 15Y + 3Z} = \frac{4x}{-2x + 12y + 3}$$

$$v' = \frac{9Y}{X + 15Y + 3Z} = \frac{9y}{-2x + 12y + 3}$$

The L^* coordinate is related to the lightness or luminance of a color, where $L^* = 0$ is black and $L^* = 100$ is white. Note that L^* for most values of Y varies as the Y to the one-third power. This variation shows that the eye does not respond in a linear fashion to a linear change in intensity. Instead, the output of a system such as a display must vary in a nonlinear fashion so as to cancel the nonlinear response of the visual system, so that the net perceived variation in intensity is linear. This process is called gamma correction. The coordinates (u', v') can be plotted as a chromaticity diagram similar to the (x, y) chromaticity diagram. The (u', v') diagram, though, is a distorted and stretched version such that the MacAdam ellipses become more round and of uniform size. In addition to this display, a color difference ΔE between two colors in the CIELUV space is given by

$$\Delta E = \sqrt{\left(L_2^* - L_1^*\right)^2 + \left(u_2^* - u_1^*\right)^2 + \left(v_2^* - v_1^*\right)^2} \qquad (2.11)$$

A value of ΔE of roughly unity represents a just noticeable difference between two colors, although the uniformity of this space is not perfect, and larger and smaller values of ΔE may correspond to being just noticeable. The CIELUV coordinates can also be expressed in cylindrical coordinates with the L^* along the axis of the cylinder, chroma being the radial coordinate defined as

$$C_{uv}^* = \sqrt{u^{*2} + v^{*2}} \qquad (2.12)$$

and hue being the azimuthal coordinate defined as

$$h_{uv} = \tan^{-1}\left(\frac{v^*}{u^*}\right) \qquad (2.13)$$

In 1976, the CIE also specified the CIELAB color space, which is a second attempt at a uniform color since a consensus could not be reached. Coordinates in the 1976 CIELAB space are designated by (L^*, a^*, b^*). The new CIELAB coordinates are calculated from the tristimulus values (X, Y, Z) through a nonlinear transformation. This transformation also takes into account the white point as set by the ambient environment. As described previously, the visual system adapts to the ambient illumination to keep white colors white. The CIELAB color coordinates are determined from the following formulas, where (X_w, Y_w, Z_w) represent the tristimulus values of the white point.

$$L^* = 116 f\left(\frac{Y}{Y_w}\right) - 16$$

$$a^* = 500\left[f\left(\frac{X}{X_w}\right) - f\left(\frac{Y}{Y_w}\right)\right] \qquad (2.14)$$

$$b^* = 200\left[f\left(\frac{Y}{Y_w}\right) - f\left(\frac{Z}{Z_w}\right)\right]$$

where $f(s) = s^{1/3}$ for $s > 0.008856$
$\quad\quad f(s) = 7.787s + 16/116$ for $s \leq 0.008856$

The CIELUV and CIELAB spaces have similar definitions for their respective L^* components, again reflecting the nonlinear perception of a linear change in intensity. The CIELAB space also incorporates color opponent processes. While the L-, M-, and S-cones respond to the red, green, and blue portions of the visible spectrum, the signals from these photoreceptors are rearranged and combined into a more efficient coding scheme. The signals are split into a luminance signal conveying brightness information and two color signals in which red and green are opposite ends of the same signal, and similarly blue and yellow form the opposite ends of a second color signal. The $a*$ coordinate determines how red or green a color is, with $a* > 0$ representing red colors and $a* < 0$ representing green colors. For the $b*$ coordinate, $b* > 0$ represents yellow colors and $b* < 0$ represents blue colors. As $a*$ and $b*$ trend toward zero, the chroma approaches zero and colors appear a more neutral gray.

As with the CIELUV color space, it is sometimes useful to define the coordinates of the CIELAB space in cylindrical coordinates. Again, the $L*$ coordinate axis lies along the axis of the cylinder and

$$C_{ab}^* = \sqrt{a^{*2} + b^{*2}} \tag{2.15}$$

is the radial coordinate representing chroma and

$$h_{ab} = \tan^{-1}\left(\frac{b^*}{a^*}\right) \tag{2.16}$$

is the azimuthal coordinate encoding hue. The color difference ΔE between two colors in the CIELAB space is

$$\Delta E = \sqrt{\left(L_2^* - L_1^*\right)^2 + \left(a_2^* - a_1^*\right)^2 + \left(b_2^* - b_1^*\right)^2} \tag{2.17}$$

In comparison to the XYZ tristimulus space, the CIELAB color space represents a far more uniform color space. In other words, the MacAdam ellipses described above become much more circular and of uniform size throughout the CIELAB space. The new coordinate system has been normalized such that a ΔE value of unity in Equation 2.17 represents a just noticeable difference between two colors.

2.4 Chromatic Aberration

The eye intrinsically has a large amount of longitudinal (axial) chromatic aberration. The eye has approximately 2.5 diopters of longitudinal

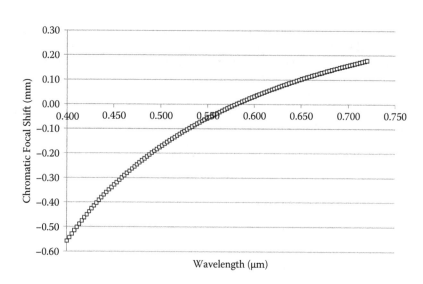

FIGURE 2.6
The chromatic focal shift of the human eye with λ = 0.580 μm assumed to correspond to best focus.

chromatic aberration.[17,18] While diopters are a common unit for visual optics, most raytracing packages describe the longitudinal chromatic aberration in terms of a focal shift. Converting the dioptric aberration to this description means that the eye has a chromatic focal shift of approximately 0.75 mm for wavelengths between 0.400 and 0.720 μm. Figure 2.6 shows the chromatic focal shift with a wavelength of 0.580 μm assumed to correspond to the eye's best focus. The value of longitudinal chromatic aberration varies modestly across the population.[19] The reason for the uniformity from eye to eye is that the chromatic aberration is due to the dispersion of the various ocular materials, which have similar properties to that of water. Effectively, the chromatic aberration of the eye is nearly equivalent to a system with a single refractive surface followed by a volume of water. The radius of curvature of this single surface matches that of the anterior surface of the cornea, and the length of this water eye matches the length of the average eyeball.

From an optical design standpoint, it is tempting to incorporate compensation for this ocular chromatic aberration into the design of the external optical system. However, attempts to correct the eye's chromatic aberration have shown no benefit. Several investigations[18,20] have created doublets that cancel most of the ocular chromatic aberration and then test visual performance. In general, there is no visual benefit. One reason for this lackluster performance is that the eye also has inherent transverse (lateral) chromatic aberration. For typical rotationally symmetric optical systems, transverse chromatic aberration does not exist on axis, and correcting chromatic aberration collapses three (or more) visible wavelengths to a single focal point on the image plane. However, since the eye is in general not rotationally symmetric, transverse chromatic aberration appears on axis (i.e., foveal vision). Correcting longitudinal chromatic aberration in this case creates three foci on the retina for red, green, and blue wavelengths. However, these focal points have a transverse separation. Consequently, any benefit to correcting the color error is lost due to the differences in magnification and image position at different wavelengths. In addition to these optical effects, the neural processes appear to compensate well for the chromatic aberration, as the anticipated color fringing effects of the aberration are typically unnoticeable. In fact, Mouroulis and Woo[21] showed that the eye can tolerate 3λ of longitudinal chromatic aberration before a noticeable deficit in visual performance is seen. When designing external optical systems that couple to the eye, the external device should be reasonably well color corrected, but the designer should not attempt to correct for chromatic aberration in the eye.

2.5 Resolution and Contrast

The resolution limit of the eye is often stated as one minute of arc. Indeed, each of the horizontal lines of the E on the 20/10 (6/3) line of a standard eye chart is separated by an angular spread of one minute of arc. This value for the resolution limit is consistent with the Rayleigh criterion. The Rayleigh criterion states that the angular separation α of two just resolved points is given by

$$\alpha = \frac{1.22\lambda}{d}\left[\frac{10800}{\pi}\right](\text{arc min}) \tag{2.18}$$

where λ is the wavelength of light and d is the pupil diameter. The factor of $10,800/\pi$ converts this result from radians into the more familiar angular measure of minutes of arc. This value approaches the limits of human resolution. However, a typical healthy eye has a visual acuity of 20/20 (6/6), and this reduction from the theoretical limit is due to ocular aberrations and scatter found in all but the most perfect eyes. Resolution is a limited metric of optical system performance. This figure of merit only considers high-contrast and high spatial frequency features in the scene. However, designers are often much more concerned about the performance of an optical system for objects that cover a broad range of contrasts and spatial frequencies. The most common metric for assessing these broad ranges of possible inputs is the modulation transfer function (MTF). The MTF considers the degradation of contrast that occurs in a sinusoidal pattern of spatial frequency, v. The contrast of a sinusoidal pattern is defined as

$$\text{Contrast} = \frac{I_{\text{max}} - I_{\text{min}}}{I_{\text{max}} + I_{\text{min}}} \tag{2.19}$$

where I_{max} is the irradiance of the peak of the sinusoid and I_{min} is the irradiance of the trough of the sinusoid. While more direct methods of measuring MTF are typically used, the following thought experiment aids in visualizing the information the MTF provides. Consider a sinusoidal target with bright bars of irradiance I_{max} and dark bars of irradiance $I_{\text{min}} = 0$. This target has a contrast of unity as determined by Equation 2.19. An optical system images this target and forms a sinusoidal image that is degraded by the aberrations and limited aperture of the optical system. Suppose the degradation causes a loss in contrast of the sinusoidal image to 0.5. MTF(v) = 0.5, or 50% in this case. This process is then repeated for a broad range of spatial frequencies. The MTF will always fall between zero and unity, with zero representing uniform illumination and unity representing the case were I_{min} falls to zero. Within a range of spatial frequencies, multiple frequencies can have MTF = 0. However, above a certain spatial frequency, the MTF will always be zero, meaning that regardless of the input, the output will always be a uniform intensity. This spatial frequency is called the cutoff frequency, v_{cutoff}.

In optical descriptions of the eye and its visual performance, the units used for describing various aspects of the system are routinely different from the units used to describe similar metrics of conventional optical systems. This difference has already appeared above in describing longitudinal chromatic aberration in units of diopters (meters^{-1}), instead of a more traditional chromatic focal shift in units of length, such as millimeters or microns. Similarly, visual acuity, which is a measure of angular resolution, was specified in units of minutes of arc, instead of radians. Spatial frequency also falls into the category of having different units for the visual description compared to the traditional optics definition. For visual systems, spatial frequency is typically measured in cycles/degree (cyc/deg). To calculate spatial frequency, the angular subtense of one period of a sinusoidal target as seen from the front nodal point of the eye is measured. Note that in this definition, the spatial frequency value is the same for both object and image space. This result is due to the properties of the nodal points where the angular subtense of an object relative to the front nodal point is the same as the angular subtense of the corresponding image relative to the rear nodal point. Most raytracing packages, on the other hand, describe spatial frequency in units of cycles/mm (cyc/mm) in the image space. Note that this image plane value of spatial frequency needs to be scaled by the system magnification to get the spatial frequency in object space. Through standardization of values based on schematic eye models of the typical human eye, a conversion exists between spatial frequency in cyc/deg and spatial frequency in cyc/mm on the retina. The standard conversion is

$$100 \text{ cyc/mm on the retina} = 30 \text{ cyc/deg} \qquad (2.20)$$

These spatial frequencies correspond to a sinusoidal target in which a single period subtends two minutes of arc, which in turn is identical to the fundamental spatial frequency of a letter on the 20/20 (6/6) line on an eye chart. This conversion also allows comparison of the cutoff frequency, v_{cutoff}, for different units. The cutoff frequency of the MTF is given by

$$v_{cutoff} = \frac{1}{\lambda(F/\#)}\left(\frac{\text{cyc}}{\text{mm}}\right) \Rightarrow \frac{d}{\lambda}\frac{\pi}{180}\left(\frac{\text{cyc}}{\text{deg}}\right) \qquad (2.21)$$

where d is the pupil diameter, λ is the wavelength, $F/\#$ is the F/number of the system, and the factor of $\pi/180$ converts from cycles/radian to cyc/deg. As an example, the cutoff frequency for a 4 mm pupil and a wavelength of $\lambda = 0.55$ μm is 423 cyc/mm or 127 cyc/deg. The maximum resolvable spatial frequency of the eye as shown above is approximately 60 cyc/deg, which corresponds to a resolution limit of one minute of arc. For physiological realizable pupil sizes of 2 to 8 mm, the cutoff frequency of the eye typically far exceeds the capabilities of the visual system to perceive the high frequencies. This limitation is due to the inherent aberrations and scatter of the ocular media, as well as the finite sample spacing between the photoreceptors in the retina. Optical designs, in turn, can ignore system performance for spatial frequencies above 60 cyc/deg (200 cyc/mm), and in general the most critical spatial frequency range will be below 30 cyc/deg (100 cyc/mm).

To specify the performance targets for an optical design, a common method is to specify the contrast (i.e., MTF value) at a given spatial frequency. The concept behind this specification now incorporates the anticipated object and detector of the entire optical system. To specify MTF performance, an understanding of the contrast and spatial frequency content of a typical object and an understanding of the ability of the detector to measure contrast and spatial frequency are required. Above, the spatial frequency measuring capabilities of the human visual system were examined. Here, the ability of the visual system to measure contrast will be demonstrated. Contrast sensitivity testing is a technique for assessing the visual system capabilities to resolve and detect low-contrast targets of different spatial frequencies. This testing technique is somewhat analogous to the conceptual MTF testing technique described above. An observer views a sinusoidal target of a given contrast. The contrast of the target is adjusted until the contrast is just at the perceptible threshold. This contrast is the minimum external target contrast required to see a target at that spatial frequency. The reciprocal of this threshold contrast is deemed the contrast sensitivity. A high value of contrast sensitivity means a low-contrast target can be detected. This process is repeated for a variety of spatial frequencies and, in general, the external contrast threshold is lower for mid-spatial frequencies than the threshold for extremely low- or high-spatial frequencies. Unlike the MTF testing, access to the image of the target formed on the retina or perceived in the brain is impossible. For this reason, contrast sensitivity testing requires feedback from the observer and has the potential to be biased by his or her observations. A variety of techniques are used to eliminate this bias, such as showing the observer multiple targets with only one of the targets actually containing a low-contrast sinusoidal pattern, while the other targets are simply uniform patches of equivalent luminance. After correcting this type of testing for randomly guessing the correct answer, the results can provide reliable measures of human contrast sensitivity.

These techniques can be taken one step further by eliminating the degradation in contrast caused by the optics of the eye. Van Nes and Bouman[22] bypassed the optics of the eye and measured contrast sensitivity directly on the retina. They used an elegant technique in which two coherent points of light are imaged into the pupil. Since the points are coherent, an interference pattern of sinusoidal fringes is created on the retina. The spatial frequency of this pattern is controlled by changing spacing between the points in the pupil. Furthermore, the contrast of the sinusoidal pattern on the retina can be adjusted by adding a third point of light in the pupil that is incoherent to the previous two points. Because of the incoherency, the third point simply adds a constant illumination superimposed onto the sinusoidal pattern, resulting in a decreased contrast pattern. Measuring the contrast thresholds of these retinal projections provides insight into the ability of the human visual system to perceive certain targets. A plot of these retinal contrast thresholds for a range of spatial frequencies is usually called the modulation threshold function. The modulation threshold function can be superimposed onto a plot of the MTF of an external optical system coupled to the eye to provide valuable information to the designer. Figure 2.7 shows an example of a modulation threshold function superimposed onto an MTF plot. The modulation threshold curve is adapted from Van Nes and Bouman. Recall that the MTF describes the amount of contrast in a sinusoidal image formed by the optical system, while the modulation threshold function is the amount of contrast required by the retina to perceive a sinusoidal target. Consequently, for spatial frequencies where the MTF exceeds the modulation threshold function, these spatial frequencies can be perceived by the visual system. When the MTF falls below the modulation threshold, the frequencies cannot be resolved by the visual system.

The Van Nes and Bouman modulation threshold data is for a 2 mm pupil and photopic conditions. These data are reasonable for the assumptions described at the outset of the chapter. Walker[23] has used the same technique in designing optical systems for visual systems. He refers to the modulation

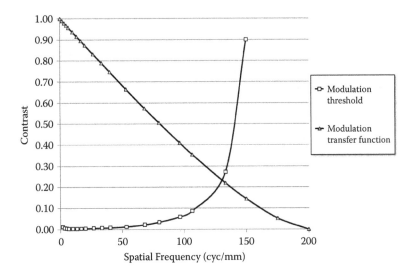

FIGURE 2.7

The modulation threshold function of Van Nes and Bouman plotted with a diffraction-limited MTF of an optical system with v_{cutoff} = 200 cyc/mm.

threshold function as the aerial image modulation (AIM) curve. Walker's values and the Van Nes and Bouman data are consistent, but differ slightly in value. The modulation threshold curve in general changes with illumination level as well, so variations are to be expected. As the light level decreases, the modulation threshold curve moves upward. The curve in Figure 2.7 is for a luminance of approximately 100 cd/m² and should be suitable for most design situations. If a system is being designed for use under dimmer conditions, then the modulation threshold function should be adjusted accordingly. Van Nes and Bouman provide modulation threshold curves for a variety of scotopic and photopic lighting conditions.[22] Tailoring the MTF performance of the optical system to meet or exceed the requirements of the visual system as defined by the appropriate modulation threshold curve ensures that the system can be used with good performance within the normal human population.

2.6 Head-Mounted Display Example

In this section, an example that incorporates some of the concepts outlined previously in the chapter will be presented. An example of a head-mounted display system will be analyzed. The display itself will be based on the OLED SXGA microdisplay described by Ghosh et al.[24] This display has 1,280 × 1,024 pixels with a pitch of 12 µm between each pixel. The dimensions of the active area of the display are 15.36 × 12.29 mm, corresponding to a diagonal size of 19.67 mm. Each pixel is made up of three rectangular subpixels with red, green, and blue filters over them. The subpixels each have a dimension of 3 × 11 µm. The chromaticity coordinates (x, y) of the red, green, and blue primaries and the white point of the microdisplay are summarized in Table 2.4. Figure 2.8 shows these primaries and the white point on a CIE chromaticity

TABLE 2.4

The CIE Chromaticity Coordinates of the White Point and the Primaries of the Microdisplay Described by Ghosh et al.

Source	CIE x	CIE y
Red	0.63	0.35
Green	0.26	0.53
Blue	0.15	0.12
White	0.30	0.37

Source: Values are adapted from Table 2 of Ghosh et al.[24]

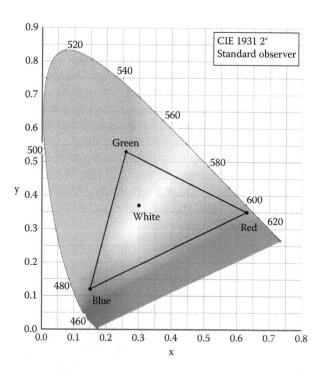

FIGURE 2.8 (See color insert.)
The gamut and white point of the microdisplay illustrated on the CIE 1931 chromaticity diagram.

diagram. The primaries represent the points of a triangular region known as the gamut within the chromaticity diagram. By mixing the outputs of the three color subpixels, any color within the triangle can be produced by the display. While the display covers a fair portion of the chromaticity diagram, the points outside of the triangle represent colors that the human visual system can perceive, but which the microdisplay is incapable of displaying.

The goal of this example is to design an optical system that allows the eye to comfortably view the display. The optical system must consist of a lens or lenses that can be molded and remain compact with an overall length from the eye to the display of less than 75 mm. The system cannot come closer to the eye than 12 mm, so as to not interfere with the eyelashes. The display should appear to have an angular size of 22.5° to the viewer. The performance of this system must allow the observer to resolve the maximum spatial frequency the display can produce, and the system/eye MTF must exceed the modulation threshold function up to this spatial frequency. The optical system must be reasonably well color corrected so as to not exceed the limits described by Mouroulis and Woo.[21] The eye itself will be assumed to be free of refractive error and have an entrance pupil diameter of 4 mm.

The first step in the design process is to perform a first-order layout with paraxial lenses representing the optical system and the eye. Since the eye is free of refractive error, the rays entering the eye need to be collimated in order to focus on the retina. Consequently, to create the head-mounted display, the microdisplay will be placed at the front focal point of an eyepiece. Light emitted from a pixel on the microdisplay will therefore be collimated when emerging from the eyepiece and can be captured and focused by the eye. Figure 2.9 shows the basic first-order layout. Based on Equation 2.2, the eye at this point has been modeled as a paraxial lens with a focal length of 16.498 mm with the retina at its back focal point. The aperture stop of the system has been located at the eye and set to a 4 mm diameter. The separation between the eye and the eyepiece is set to 15 mm. This value arises from the minimum eye relief of 12 mm plus 3 mm for the separation between the corneal vertex and the entrance pupil found in the real eye (see VE in Table 2.2). When the paraxial eye model is replaced with a thick-lens version, this separation will be reduced back to 12 mm. The field height on the object

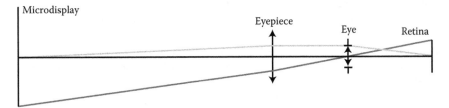

FIGURE 2.9
First-order layout of a head-mounted display system.

TABLE 2.5

Paraxial Layout and Raytrace of Head-Mounted Display System

Surface Number	Surface	Focal Length (mm)	Thickness (mm)	y (mm)	u'	$\bar{y}$ (mm)	$\bar{u}'$
0	Microdisplay		47.177	0.000	0.042	−9.385	0.136
1	Eyepiece (paraxial)	47.177	15.000	2.000	0.000	−2.984	0.199
2	Eye (paraxial)	16.498	16.498	2.000	−0.121	0.000	0.199
3	Retina			0.000		3.282	

plane is set to 9.385 mm, half the diagonal dimension of the microdisplay. The paraxial lens powers, separations, and properties of the marginal and chief rays are summarized in Table 2.5. Note that the focal length of the eyepiece, $f_{eyepiece}$, has been chosen to meet the system length criteria of less than 75 mm from the eye to the display. This focal length was also chosen to provide the desired field of view with the chief ray angle $\bar{u}_1'$ entering the eye as

$$\bar{u}_1' = \tan\left(\frac{22.5°}{2}\right) = 0.199 \tag{2.22}$$

The first-order layout also shows that the paraxial magnification, m, of the system is

$$m = \frac{3.282}{-9.385} = \frac{-f_{eye}}{f_{eyepiece}} = -0.35 \tag{2.23}$$

In other words, the paraxial magnification is proportional to the ratio of the focal length of the eye to the focal length of the eyepiece. Finally, the first-order layout shows that the clear aperture of the eyepiece is

$$\text{Clear Aperture} = 2\left(|y_1| + |\bar{y}_1|\right) = 2(2 + 2.984) = 9.968 \text{ mm} \tag{2.24}$$

Based on these results, the eyepiece will approximately be $F/4.7$ and meet the system length and field of view requirements.

Next, the maximum spatial frequency that can be displayed needs to be analyzed. Define $v_{max,o}$ as this maximum spatial frequency, where the o in the subscript denotes that this value is measured in the object plane. The value of $v_{max,o}$ corresponds to the case where adjacent pixels on the microdisplay alternate between dark and bright emissions. This scenario means that one cycle of a sinusoidal pattern (albeit binary in this case) can be displayed with two pixels. This maximum spatial frequency is then given by

$$V_{max,o} = \frac{1 \text{ cycle}}{2(\text{pixel width})} = \frac{1 \text{ cycle}}{0.024 \text{ mm}} = 41.67 \frac{\text{cyc}}{\text{mm}} \quad (2.25)$$

The paraxial magnification from Equation 2.15 shows how the size of this pattern changes as it is imaged onto the retina. The maximum spatial frequency on the retina that can be displayed is

$$V_{max,i} = |m| V_{max,o} = 14.57 \frac{\text{cyc}}{\text{mm}} \quad (2.26)$$

where the i in the subscript now denotes the value being measured in the image plane. Note that this spatial frequency is well below the 100 cyc/mm or so limit that the eye is capable of detecting, as described above. In other words, the display and *not* the eye is the limiting factor for the targets that can be produced with this head-mounted display. It is also interesting to look at Equations 2.25 and 2.26 to examine how such a display system could be pushed toward the limits of human vision. One way of increasing $V_{max,i}$ is to increase the absolute value of the paraxial magnification, which requires the focal length of the eyepiece, $f_{eyepiece}$, to be shortened, as can be seen in Equation 2.23. The second way of increasing $V_{max,i}$ is to decrease the pixel pitch, as seen in Equation 2.25. Thus, faster optics and smaller pixels will allow the maximum displayable spatial frequency to increase to match the performance of the human visual system.

Continuing with the display design process, the paraxial eyepiece lens will be replaced with a thick lens. To be moldable and to reduce chromatic effects, Zeonex will be used to form a singlet for the eyepiece. To keep the system light, the thickness of the lens will be set to 2 mm. The eye in this case is still modeled as a paraxial lens. By maintaining this state, the chromatic aberration inherent to the eyepiece can be distinguished from the chromatic aberration intrinsic to the eye. The singlet eyepiece system is shown in Figure 2.10, and the design parameters are summarized in Table 2.6. Problems with the singlet design are immediately visible from the figure. As to be expected

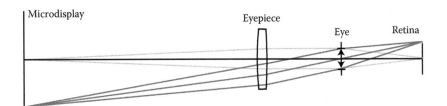

FIGURE 2.10
Head-mounted display system with a singlet eyepiece.

TABLE 2.6

Singlet Lens Eyepiece Layout of Head-Mounted Display System

Surface Number	Surface	Radius (mm)/ Focal Length (mm)	Thickness (mm)	Material
0	Microdisplay		47.177	Air
1	Singlet front	29.675 (radius)	2.000	Zeonex
2	Singlet rear	−181.365 (radius)	15.000	Air
3	Eye (paraxial)	16.498 (focal length)	16.498	Air
4	Retina			

from a singlet, the eyepiece introduces longitudinal chromatic aberration on the axis. Through raytracing, this longitudinal chromatic aberration corresponds to about 1λ of error. Based on Mouroulis and Woo's result[21] this level is probably tolerable, but a much more severe error arises in the setup. Since the eye pupil is serving as the system aperture, the bundle of rays coming from the corner of the display passes through a peripheral portion of the eyepiece. This portion of the lens can be thought of as a wedge-shaped lens, and consequently, it also acts like a prism and disperses the incident light. The singlet eyepiece system has severe transverse chromatic aberration, which becomes quickly apparent to the observer moving off the optical axis. This result suggests that a second lens is needed for the eyepiece that can compensate for these chromatic effects.

To improve the chromatic properties of the head-mounted display, the singlet in the previous step is replaced by an air-spaced doublet. One lens is made of Zeonex, while the other is made of polycarbonate. The difference in the material dispersions provides the flexibility to dramatically reduce both the longitudinal and transverse chromatic aberrations of the previous design. To maintain the apparent field of view of the microdisplay with respect to the eye and to account for the thickness of the eyepiece lenses, the separation between the microdisplay and the eyepiece is reduced. The parameters of this new design are summarized in Table 2.7. The air-spaced

TABLE 2.7

Air-Spaced Doublet Eyepiece Layout of Head-Mounted Display System

Surface Number	Surface	Radius (mm)/Focal Length (mm)	Thickness (mm)	Material
0	Microdisplay		44.000	Air
1	Element 1 front	27.865 (radius)	1.000	Polycarbonate
2	Element 1 rear	11.265 (radius)	1.000	Air
3	Element 2 front	11.640 (radius)	3.000	Zeonex
4	Element 2 rear	−103.998 (radius)	15.000	Air
5	Eye (paraxial)	16.498 (focal length)	16.498	Air
6	Retina			

doublet design now has less than 0.25λ of chromatic aberration across the entire field. The on- and off-axis performance of this system are good, with only some residual astigmatism in the off-axis case. Finally, the paraxial model of the eye can be replaced with the Arizona eye model, summarized in Table 2.1. The separation between the eyepiece and the cornea of the eye model is reduced to 12 mm. This reduced distance meets the specification for the eye relief, while at the same time places the entrance pupil of the Arizona eye model at the stop location from the previous designs that used the paraxial eye model. Figure 2.11 shows the final head-mounted display system with the eye. With the schematic eye model in place, the MTF of the combined eyepiece–eye system can be calculated to determine if the MTF values exceed the modulation threshold function for the desired spatial frequencies. Figure 2.12 shows the MTFs for the on-axis case and at the corner

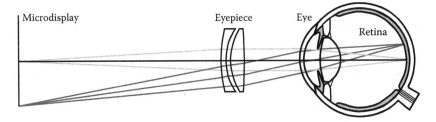

FIGURE 2.11
Head-mounted display system with an air-spaced doublet eyepiece.

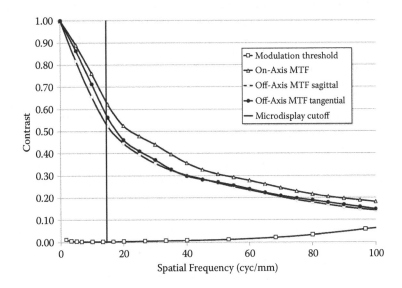

FIGURE 2.12
MTFs for both on- and off-axis cases for the final head-mounted display system. The modulation threshold curve is shown as well.

of the display. Note that the MTFs of the system with the paraxial eye model were nearly diffraction limited and well corrected for color. In Figure 2.12, the reduction in the MTFs is due to the intrinsic aberrations, both monochromatic and chromatic, of the eye. As described previously, the visual system appears well suited for coping with the chromatic aberration of the eye, so this mechanism is exploited in the design of the external device. The MTFs for the on- and off-axis cases are similar, suggesting good performance over the entire display. Also shown in Figure 2.12 is the modulation threshold curve. From Equation 2.18, the maximum spatial frequency on the retina is 14.57 cyc/mm. The MTFs at this frequency are between 0.55 and 0.65, which greatly exceeds the modulation threshold. Consequently, no difficulties in resolving the frequencies displayed by the microdisplay are expected.

2.7 Summary

This chapter explored the functions of the human eye and visual system. In many optical systems, the eye becomes the ultimate detector, and understanding its properties and limitations is required to successfully design such systems. Microscopes are one such example of an optical system with the eye as the final detector. Historically, microscopes were primarily used with biological and inert specimens below the limits on human acuity. While high-resolution cameras and displays have largely replaced the eye for these systems, eye-based microscopes are still widely used for surgery. Cameras and displays still cannot provide the surgeon with the quality and resolution needed to perform many of today's intricate surgical procedures, so designing such systems to work in conjunction with the eye's properties allows surgeons to push the boundaries and explore new techniques. A second example of an optical system that uses the eye as the final detector is the head-mounted display, such as the design example presented in this chapter. The applications of this type of device are widespread. In the entertainment arena, head-mounted displays can provide an immersive gaming environment, especially if the display can cover large fields of view. This allows the participant to step inside the game, adding new dimensions and reality to the play. Such a system has application to security as well. These immersive environments can provide training to security personnel by allowing them to enter different scenarios and respond accordingly. Head-mounted displays are also useful for the soldier in the field, as they can provide augmented reality in which information is overlaid onto the scene viewed by the soldier. Information can be simple, such as coordinates and heading information obtained from a global positioning device, or more complex, such as sensor output, automated threat detection, and night vision. Systems that use the eye as a final detector have a common thread. In general, these systems take information that is typically undetectable or not resolvable by the human visual system and remap them to an input in which they can be imaged onto the retina and processed by the visual system. The advantage of such a system is that it leverages a wider array of physical properties from the outside world and then couples it to the visual system, which often far surpasses artificial image processing systems in detection, reaction time, and complexity of recognized objects.

To take full advantage of optical systems coupled to the eye and visual system, an understanding of its capabilities and limitations is needed. In this chapter, details regarding the limitations to resolution, aberration, and contrast were outlined. In general, providing performance in an optical system that exceeds a spatial frequency of 100 to 200 cyc/mm on the retina is unnecessary. Furthermore, correcting chromatic aberration in the external optical system is beneficial, but attempting to further correct chromatic aberration inherent to the eye has not shown an improvement to performance. Finally, the modulation threshold or minimum contrast required by the retina for detection was seen to be below 1% for most spatial frequencies, but this requirement rapidly rises for high spatial frequencies approaching 100 cyc/mm. Providing sufficient contrast to the retina is essential for detection of targets.

As a final caveat for this chapter, the values and dimensions provided here are for a normal or average eye. As with other human features, such as height, weight, hair color, and body type, there is a large variation in the population as to the shape of the ocular components and the performance of the visual system. Furthermore, additional variables, such as age, scatter, uncorrected refractive error, ocular disease, and neural issues, have not been taken into account here. However, while there will always be individuals who are outliers, targeting the values provided here in the optical design process should have broad applicability.

References

1. Le Grand, Y., and El Hage, S. G. 1980. *Physiological optics*. Berlin: Springer-Verlag.
2. Lotmar, W. 1971. Theoretical eye model with aspheric surfaces. *J Opt Soc Am* 61:1522–29.
3. Navarro, R., Santamaría, J., and Bescós, J. 1985. Accommodation-dependent model of the human eye with aspherics. *J Opt Soc Am A* 2:1273–81.
4. Liou, H.-L., and Brennan, N. A. 1997. Anatomically accurate, finite schematic model eye for optical modeling. *J Opt Soc Am A* 14:1684–95.
5. Schwiegerling, J. 2004. *Field guide to visual and ophthalmic optics*. Bellingham, WA: SPIE Press.

6. CIE. 1926. *Commission Internationale de l'Eclairage Proceedings, 1924.* Cambridge: Cambridge University Press.

7. Judd, D. B. 1951. Report of U.S. Secretariat Committee on Colorimetry and Artificial Daylight. In *Proceedings of the Twelfth Session of the CIE*, Stockholm, 11. Vol. 1. Paris: Bureau Central de la CIE.

8. Sharpe, L. T., Stockman, A., Jagla, W., and Jägle, H. 2005. A luminous efficiency function, V*(λ), for daylight adaptation. *J Vision* 5:948–68.

9. CIE. 2007. *Fundamental chromaticity diagram with physiological axes.* Parts 1 and 2, Technical Report 170-1. Vienna: Central Bureau of the CIE.

10. Malacara, D. 2002. *Color vision and colorimetry: Theory and applications.* Bellingham, WA: SPIE Press.

11. Wright, W. D. 1928. A re-determination of the trichromatic coefficients of the spectral colours. *Trans Opt Soc* 30:141–64.

12. Guild, J. 1931. The colorimetric properties of the spectrum. *Phil Trans Royal Soc London A* 230:149–87.

13. CIE. 1932. *Commission Internationale de l'Éclairage Proceedings, 1931.* Cambridge: Cambridge University Press.

14. CIE. 1964. *Vienna Session, 1963*, 209–20. Committee Report E-1.4.1, Vol. B. Paris: Bureau Central de la CIE.

15. Vos, J. J. 1978. Colorimetric and photometric properties of a 2-deg fundamental observer. *Color Res Application* 3:125–28.

16. MacAdam, D. L. 1942. Visual sensitivities to color differences in daylight. *J Opt Soc Am* 32:247–74.

17. Wald, G., and Griffin, D. R. 1947. The change in refractive power of the human eye in dim and bright light. *J Opt Soc Am* 37:321–36.

18. Bedford, R. E., and Wyszecki, G. 1957. Axial chromatic aberration of the human eye. *J Opt Soc Am* 47:564–65.

19. Thibos, L. N., Ye, M., Zhang, X., and Bradley, A. 1992. The chromatic eye: A new reduced-eye model of ocular chromatic aberrations in humans. *Appl Opt* 31:3594–600.

20. van Heel, A. C. S. 1946. Correcting the spherical and chromatic aberrations of the eye. *J Opt Soc Am* 36:237–39.

21. Mouroulis, P., and Woo, G. C. 1988. Chromatic aberration and accommodation in visual instruments. *Optik* 80:161–66.

22. Van Nes, F. L., and Bouman, M. A. 1967. Spatial modulation transfer in the human eye. *J Opt Soc Am* 57:401–6.

23. Walker, B. H. 2000. *Optical design for visual systems.* Bellingham, WA: SPIE Press.

24. Ghosh, A. P., Ali, T. A., Khayrullin, I., et al. 2009. Recent advances in small molecule OLED-on-silicon microdisplays. *Proc SPIE* 7415:74150Q-1–12.

3

Stray Light Control for Molded Optics

Eric Fest

CONTENTS

71

3.1 Introduction

Stray light is generally defined as unwanted light that reaches the focal plane of an optical system. A classic stray light problem occurs when photographing a scene in which the sun is just outside the field of view (FOV). Light from the sun strikes the camera and, by mechanisms such as ghost reflections, surface roughness scatter, and aperture diffraction, reaches the focal plane and creates bright spots that can obscure the subject. An example of this phenomenon is shown in Figure 3.1.

FIGURE 3.1
Stray light in a daytime photograph due to the sun just outside the field of view (FOV). Multiple mechanisms are responsible for the stray light in this image: ghost reflection artifacts can be seen in both circled regions, and the region on the left also contains artifacts from aperture diffraction and surface roughness and particulate contaminaton scattering.

Another common stray light problem occurs when photographing a scene at night in which there are one or more sources of light that are bright and have a narrow extent in FOV, such as street lamps. The same stray light mechanisms that caused the bright spots in Figure 3.1 can cause similar bright spots in this scenario, as shown in Figure 3.2. In both cases, stray light in the optical system has resulted in unwanted light in the final image.

The consequences of this unwanted light depend on the purpose for which this optical system was constructed; in a consumer photography application (such as a low-cost digital camera), stray light may be a minor annoyance that requires the user to take the picture again from a different angle. In a security or military application, this stray light may obscure the intended target, resulting in a total failure of the system. Understanding the consequences of stray light prior to designing the system ensures that the appropriate steps can be taken to control it.

For those readers who are not interested in understanding the details of stray light analysis and control, the following is a list of best practices used to control stray light. Be warned: This list is not a substitute for analyzing the stray light performance of an optical system, and unexpected results may occur from its use:

FIGURE 3.2
Stray light in a nighttime photograph of bright sources of narrow angular extent. Multiple mechanisms are responsible for the stray light in this image: ghost reflections are responsible for the circled artifacts, and surface roughness and particulate contamination scattering are responsible for the enlargement or "blooming" of the streetlights. The star-shaped patterns coming from the streetlights are due to aperture diffraction.

- Whenever possible, use a cylindrical baffle in front of the first element to shadow the system from illumination by off-axis stray light sources.

- Block direct illumination of the detector (such as that which can occur in a molded optic system with uncovered flanges; see Figure 3.3) with baffles.

- Apply Anti-Reflection (AR) coatings to all refractive optical surfaces.

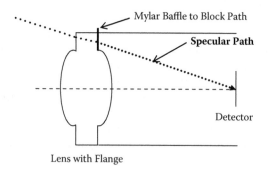

FIGURE 3.3
Direct-illumination stray light path (zero-order) in a molded optic system with uncovered flanges.

- Anodize, paint black, or roughen all surfaces near the optics, especially the inner diameter of lens barrels and struts.
- Paint black or roughen the edges of lenses. If the lens has a flange that is used for mounting or is left from the molding process, paint or roughen it as well. Exposed faces of flanges should be roughed, painted, or covered with a baffle (Mylar can be used for this purpose).
- Make the Root Mean Square (RMS) surface roughness of the optical surfaces as low as possible.
- Keep the optical surfaces as clean (i.e., as free of particulate and molecular contamination) as possible.
- Whenever possible, use optical designs with field stops.

This chapter is broadly divided into two sections. The first section presents the background information necessary to perform stray light analysis: basic terminology, radiometry, and a discussion of stray light mechanisms such as ghost reflections and surface scatter. The second section will focus on the application of these concepts in the development of an optical system, and will discuss options for controlling stray light in the optical design, basic baffle design, and other stray light control methods. A process will be introduced whereby requirements for the stray light performance of the optical system are established and then flow down to the optical and baffle design of the system. The system is then analyzed and tested to ensure that the requirements are met. This will result in an optical system whose stray light control is sufficient to meet the needs of its intended users, which is the primary goal in any stray light analysis effort.

3.2 Stray Light Terminology

3.2.1 Stray Light Paths

A stray light path is a unique sequence of events experienced by a beam of light, ending by absorption at the image plane. An example of a stray light path description is: "Light leaves the sun, transmits through lens 1, ghost reflects off of lens 2, ghost reflects off of lens 1, transmits through lens 2, and hits the detector" (see the next section for the definition of ghost reflections). Any optical system has many such paths. Paths are often categorized by their order, which refers to the number of stray light mechanisms (or events) that occur in the path. For instance, the path described above is a second-order path, because it contains two ghost reflection events. Non-stray light events in the path (such as "transmits through lens 1") are not considered in the order count. As will be discussed in Section 3.4.1.1, it's possible for a stray light path to have zero order (direct illumination of the detector). In most systems, any path of order greater than two does not usually produce a significant amount of light on the focal plane, since the magnitude of this light usually varies as the nth power, where n is the order.

3.2.2 Specular and Scattered Stray Light Mechanisms

Stray light mechanisms generally fall into one of two categories: specular or scatter. The difference between the two is that light from a specular mechanism obeys Snell's laws of reflection and refraction; that is, the angle of all reflected rays relative to the surface normal of the reflecting surface is equal to the angle of incidence, and the angle of refraction θ' relative to the surface normal of the refracting surface of all refracted rays is given by

$$n \sin\theta = n' \sin\theta' \tag{3.1}$$

where n and θ are the refractive index of the incident media and the angle of the incident ray relative to the surface normal of the refracting surface, respectively, and n' is the refractive index of the transmitting media. The difference between these mechanisms is illustrated in Figure 3.4. An example of a specular mechanism is a ghost reflection. Scatter mechanisms, by contrast, do not obey Snell's laws, and the angle of the scattered ray with respect to the surface normal can take on any value. An example of a scatter mechanism is scattering from optical surface roughness. Light never undergoes a perfect specular reflection or transmission through a surface, even a highly polished optical surface; there is always at least some small amount of scatter. The decision about whether or not to model scattering from a surface depends on the magnitude of the scatter and the sensitivity of the optical system being analyzed. In general, when performing a stray light analysis, it is better to be

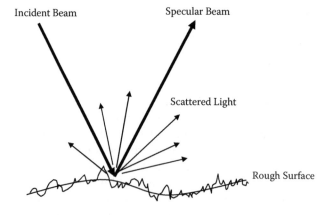

FIGURE 3.4
Scattered and specular rays reflected from a rough surface.

conservative and model the scatter in the system, even scatter from optical surfaces. The effect on the stray light performance of the system due to this scatter can then be evaluated.

3.2.3 Critical and Illuminated Surfaces

A critical surface is one that can be seen by the detector (either an electronic detector or the human eye), and an illuminated surface is one that is illuminated by a stray light source. In order for stray light to reach the focal plane, there must be at least one surface that is both critical and illuminated. This concept is illustrated in Figure 3.5, and is central to the process of stray light analysis and design. In general, a stray light mechanism must occur on a surface that is both critical and illuminated in order to have any impact on the stray light performance of the system.

3.2.4 In-field and Out-of-Field Stray Light

Sources of stray light can be either inside or outside the nominal FOV of the system, and the stray light that results from these sources is referred to as in-field or out-of-field stray light, respectively. The stray light shown in Figure 3.1 is out-of-field stray light, and the stray light shown in Figure 3.2 is in-field.

3.2.5 Sequential and Nonsequential Raytracing Programs

Detailed design of optical systems is performed using two types of commercially available programs: sequential and nonsequential geometric raytracing programs. In a sequential raytracing program, the order in which rays intersect surfaces is nominally limited to a single path, namely, the primary optical path. Typically, the sequential program is used first to optimize the image quality of the optical system. The most basic design decisions about the system (such as the number of elements, their shape, etc.) are typically made using these programs, examples of which include CODE V[1]

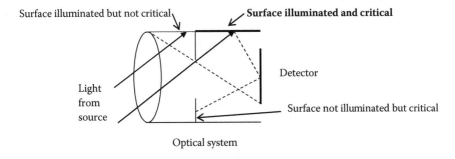

FIGURE 3.5
Illuminated and critical surfaces.

and ZEMAX.[2] After the image quality of the system has been optimized, the stray light performance of the system is designed and analyzed using a non-sequential program, in which the order of surfaces encountered by each ray is determined only by the direction of the ray and the position and size of the surfaces. In such programs, rays are free to take any path, as long as it is physically possible to do so. Examples of nonsequential programs include FRED,[3] ASAP,[4] and TracePro.[5] Most sequential programs feature nonsequential raytracing modes, and vice versa; however, that is not their primary purpose.

3.3 Basic Radiometry

Stray light analysis cannot be understood without understanding radiometry, which is the science of quantifying the "brightness" of optical sources and how this brightness propagates through an optical system to its focal plane. The term *source* can be used to refer to any object from which light is radiating, regardless of whether or not the light is generated by the object or reflected or transmitted by it. Radiometry can be used to quickly estimate the stray light performance of a system, without the labor-intensive process of building a stray light model in a raytracing analysis program. Understanding radiometry is also important because the results of the simple radiometric analysis can be used to roughly validate the results of a more detailed raytracing analysis. Setting up a detailed raytracing analysis can be very complicated, and it is easy to make errors in setting up this analysis; therefore, a first-order radiometric analysis can be used to roughly validate some of the results. A comprehensive review of radiometry is beyond the scope of this book; however, there are a number of good references.[6–8]

3.3.1 Basic Radiometric Terms

3.3.1.1 *Flux or Power*

The flux, or power, of a source is equal to the number of photons/s it outputs. In this chapter, this quantity will be represented by the symbol Φ. Flux will be expressed in units of photons/s in this book because these units make the analysis of digital camera systems easier, since the relationship between the number of electrons (and thus the photocurrent, which is equal to electrons/s, or amperes) induced in a detector and the number of photons incident on it is related by the quantum efficiency of the detector η:

$$(\text{\# of electrons induced in the detector}) =$$

$$\eta(\text{\# of photons incident on the detector})$$

(3.2)

Watts and lumens are two other units that are also often used to represent flux. The number of watts output by an optical source is related to the number of photons/s it outputs by the equation

$$\text{Watts} = \sum_{i=1}^{n} \frac{hc}{\lambda_i} \qquad (3.3)$$

where n is the number of photons, h is Planck's constant (6.626E-34 Joule*s), c is the speed of light (3E8 m/s), and λ_i is the wavelength (in meters) of the ith photon. Watts are used in the analysis of all types of optical systems, particularly infrared (IR) systems using microbolometer array detectors. Lumens (lm) are photometric units, which are useful in the analysis of visible light systems, and can be computed from the flux (in watts) by the equation

$$\text{Lumens} = 680 \int \Phi(\lambda) p(\lambda) d\lambda \qquad (3.4)$$

where $p(\lambda)$ is the photopic response curve, which quantifies the response of a standard human eye to light. The numeric values of this curve are available from a variety of sources online. Plots of photopic response curves were shown in Figure 2.2. The integral is evaluated over the visible spectrum (roughly 0.3 microns to 0.7 microns). Nearly all of the radiometric units discussed below have photometric equivalents, and these equivalents will be introduced here as well. Equations in this section that contain Φ are valid only if Φ does not vary over the given area, solid angle, or projected solid angle.

3.3.1.2 Exitance

The exitance M of a source is equal to

$$M = \frac{\Phi}{A} \qquad (3.5)$$

where Φ is the power emitted by the source and A is the area of the source. Exitance is specified in photons/s/unit area, or, in photometric units, in Lux (lm/m²).

3.3.1.3 Solid Angle

In the spherical coordinate system shown in Figure 3.6, the solid angle ω of an object as viewed from a particular point in space is equal to

$$\omega = \int_{\phi_1}^{\phi_2} \int_{\theta_1}^{\theta_2} \sin(\theta) d\theta \qquad (3.6)$$

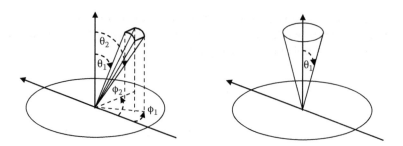

FIGURE 3.6
Angles used in the definition of a solid and projected solid angle (left) and in a right angle cone (right).

where ϕ_1 and ϕ_2 define the extent of the object in the azimuthal coordinate and θ_1 and θ_2 define the extent of the object in the elevation coordinate. The units of solid angle are steradians (sr).

A geometry often encountered in radiometry is also shown in Figure 3.6, in which $\theta_1 = 0$, $\phi_1 = 0$, and $\phi_2 = 2\pi$. This geometry is called a right angle cone, and its solid angle is equal to

$$\omega = 2\pi\left[1 - \cos(\theta_2)\right] \tag{3.7}$$

3.3.1.4 Intensity

The intensity *I* of a point source is given by

$$I = \frac{\Phi}{\omega} \tag{3.8}$$

where ω is the solid angle the point source is emitting into. Intensity can only be defined for point sources, that is, sources that have an infinitely small extent. Though no real-world sources exactly meet these criteria, this way of defining the brightness of a source is often useful. Intensity is specified in photons/s-sr or, in photometric units, in candela (lm/sr).

3.3.1.5 Projected Solid Angle

The definition of projected solid angle is similar to the definition of solid angle, except for the addition of a cosine term:

$$\Omega = \int_{\phi_1}^{\phi_2}\int_{\theta_1}^{\theta_2} \sin(\theta)\cos(\theta)d\theta \tag{3.9}$$

The units of projected solid angle are steradians, just as for solid angle.

There are a number of common cases for which the value of the projected solid angle is simple to compute. The first of these is the right angle cone (considered above), which is equal to

$$\Omega = \pi \sin^2(\theta_2) \tag{3.10}$$

The projected solid angle of an object can also be approximated as

$$\Omega \simeq \frac{A}{d^2} \tag{3.11}$$

where A is the area of the object and d is the distance between the object and the observation point. This approximation is valid when $d \gg A$.

The projected solid angle of an optical system is related to its F-number ($F/\#$) by the equation

$$\Omega = \frac{\pi}{4(F/\#)^2} \tag{3.12}$$

3.3.1.6 Radiance

The radiance of a source L is equal to

$$L = \frac{\Phi}{A\Omega} \tag{3.13}$$

where Φ is the power emitted by the source, A is the area of the source, and Ω is the projected solid angle that the source is emitting into. The units of radiance are photons/s-unit area/sr, or, in photometric units, candela/m² (also called nits). As will be shown, radiance is a very important quantity in radiometric analysis, as it is often used as the most complete description of the brightness of a source.

3.3.1.6.1 Blackbody Radiance

The Planck blackbody equation can be used to compute the radiance of an extended source from its temperature:

$$L = \int_{\lambda_1}^{\lambda_2} \frac{C_1}{\lambda^4 \left[\exp(C_2/\lambda T) - 1 \right]} d\lambda \tag{3.14}$$

where λ_1 and λ_2 are the minimum and maximum wavelengths of the waveband of interest (in μm), $C_1 = 5.99584\text{E}+22$ photons-μm⁵/s-cm²,

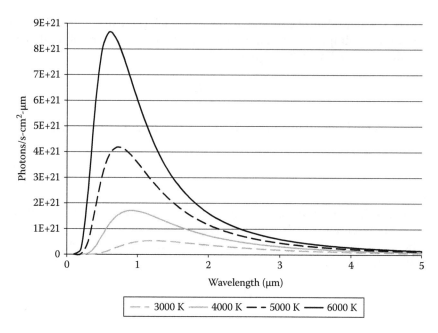

FIGURE 3.7
Radiance vs. wavelength for ideal blackbodies.

$C_2 = 14387.9$ μm-K, and T is the temperature of the source in Kelvin. There is no simple closed form of this integral, and it must be evaluated numerically. The value of integrand vs. wavelength is plotted for several temperatures in Figure 3.7. This equation is accurate for thermal sources of light such as the sun ($T \sim 5{,}900$ K) and other sources that are much hotter than their surroundings. This equation is less accurate for man-made sources of light such as lightbulbs, and in these cases it is recommended that the manufacturer's specifications for the output of the source be used to model its radiance.

3.3.1.7 Irradiance

The irradiance E incident on a surface is equal to

$$E = \frac{\Phi}{A} \tag{3.15}$$

where Φ is the power incident on the surface and A is the area of the surface. The units of irradiance are photons/s-unit area or, in photometric units, lux (lm/m²). The only difference between exitance and irradiance is the direction of propagation of light.

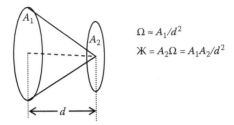

$$\Omega \approx A_1/d^2$$

$$\text{Ж} = A_2\Omega = A_1A_2/d^2$$

FIGURE 3.8
Definition of throughput.

3.3.1.8 Throughput

The throughput ж of an optical system is equal to $A\Omega$, where A is the area of the surface upon which light is incident and Ω is the projected solid angle of the source as viewed from the surface, as shown in Figure 3.8. The units of throughput are unit-area-sr. This quantity (also called the A-omega product) is fundamental in understanding how optical power propagates from a source to a surface, since its value is the same at any point within an optical system.

3.3.1.9 Bidirectional Scattering Distribution Function (BSDF)

BSDF is a means of quantifying the brightness of a scattering surface, and is equal to

$$\text{BSDF} = \frac{L}{E} \tag{3.16}$$

where L is the radiance of the scattering surface and E is the irradiance incident on it. The units of BSDF are 1/sr. BSDF is often referred to as either the bidirectional reflectance distribution function (BRDF) or bidirectional transmittance distribution function (BTDF), depending on the direction the scattered light is propagating relative to the scattering surface. The BSDF of most real-world surfaces varies as a function of many parameters, most importantly wavelength, angle of incidence (AOI = θ_i) relative to the surface normal, and the scatter angle (θ_s) relative to the surface normal. These angles are shown in Figure 3.9.

The ratio of the power incident on a scattering surface to the total amount of power scattered by it in the reflected or transmitted direction is called the total integrated scatter (TIS), and is equal to the integral of the BSDF over the projected solid angle of the hemisphere:

$$\text{TIS} = \int_0^{2\pi}\int_0^{\pi/2} \text{BSDF}\sin(\theta_s)\cos(\theta_s)d\theta_s d\phi \tag{3.17}$$

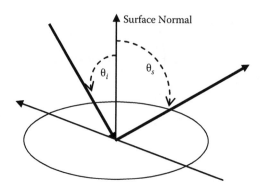

FIGURE 3.9
Angle of incidence (θ_i) and scatter angle (θ_s).

When comparing the magnitude of different scattering mechanisms, it is often useful to compare the TIS of the mechanisms rather than the BSDF, since it is often easier to compute the TIS. This concept will be discussed further in Section 3.4.2.

3.3.1.10 Putting It All Together—Basic Radiometric Analysis

The radiometric terms defined above will now be used to compute the transfer of optical power in a number of simple scenarios.

The irradiance E on a surface due to illumination by a point source of intensity I is equal to

$$E = \frac{I\cos^3(\theta)}{d^2} \tag{3.18}$$

where d is the distance between the point source and the surface and θ is the angle between the vector to the source and the surface normal, as illustrated in Figure 3.10.

The irradiance E on a plane due to illumination by an extended source of radiance L, area A, distance d, and angle θ is given by

$$E = L\frac{A}{d^2}\cos^4(\theta) = L\Omega\cos^4(\theta) \tag{3.19}$$

where Ω is the projected solid angle of the source as seen from the point where E is being computed, as shown in Figure 3.10.

Equation 3.19 can also be used to compute the irradiance on the focal plane of an optical system due to an object of radiance L in the center of its field of view (FOV), except Ω is given by the solid angle of the optical system (as computed using Equation 3.12):

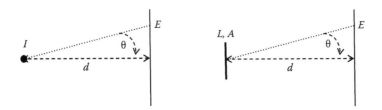

FIGURE 3.10
Irradiance from a point source on a plane (left) and from an extended source on a plane (right).

$$E = L\left[\frac{\pi}{4(F/\#)^2}\right]\tau \tag{3.20}$$

where τ is the transmittance of the optical system.

The final calculation to be considered here is the irradiance E, in the center of the focal plane of an optical system due to scattering by one or more of the surfaces in the optical path, which is given by

$$E = L\Omega\cos(\theta)(BSDF)\left[\frac{\pi}{4(F/\#)^2}\right]\tau \tag{3.21}$$

where L is the radiance of the extended source illuminating the scattering surface, Ω is the solid angle of the extended source as viewed from the scattering surface, θ is the angle of the source from the optic axis, BSDF is the BSDF of the scattering surface, and τ is the transmittance of the optical system, as shown in Figure 3.11. This equation assumes that the scattering surface is flat, and therefore the more curved the scattering surface actually

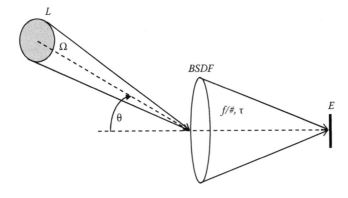

FIGURE 3.11
Irradiance on the focal plane due to scattering in an optical system.

is, the less accurate this equation. This illustrates the need to use raytracing software in order to obtain more accurate results in which surface curvatures (among other things) are accounted for. Despite this approximation, this equation is very useful, and can be used to make a quick estimate of the stray light performance of a system without the need to model it in a raytracing program.

3.4 Basic Stray Light Mechanisms

Stray light mechanisms are the physical processes experienced by a beam of light that direct it to the focal plane via a path other than the primary optical path. These mechanisms can be divided into two categories: specular and scattering. A list of common surface types used in optical systems and the stray light mechanisms that can occur on each are given in Table 3.1. Note that both specular and scattered mechanisms can occur on some surface types, such as refractive optical surfaces.

3.4.1 Specular Mechanisms

Before discussing specular mechanisms, note that nearly all of them must be modeled using a nonsequential raytracing program.

3.4.1.1 Direct Illumination of the Detector

This is perhaps the most serious of all of stray light mechanisms, and occurs when the detector of an optical system is directly illuminated by a source via an optical path other than the primary imaging path. The classic example

TABLE 3.1

Common Surface Types in Optical Systems and Their Associated Stray Light Mechanisms

Surface Type	Specular Mechanisms	Scattering Mechanisms
Refractive optical surface (lens, filter, or window)	Ghost reflections	Optical surface roughness, haze, and contamination scatter
Reflective optical surface (mirror)	Specular mirror	Optical surface roughness and contamination scatter
Diffractive optical surface	None	Scattered orders, scattering from transition zones
Metal or plastic (painted or unpainted) mechanical surface	Specular reflection	Mechanical surface roughness scatter

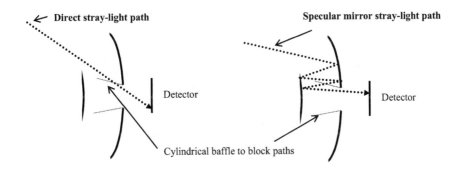

FIGURE 3.12
Direct (left) and specular mirror (right) stray-light paths in a Cassegrain telescope.

of this mechanism occurs in a Cassegrain telescope in which the detector can be directly illuminated through the hole in the primary mirror, as shown in Figure 3.12. In this case, there is no scattering (and hence no BSDF model) involved: the source itself is the critical and illuminated surface. Obviously, this mechanism is not an issue for every type of optical system: most consumer camera systems are all refractive and have no such stray light path. However, for reflective systems such as the Cassegrain telescope, these paths can be very serious and need to be identified and dealt with early on in the design of the system. The irradiance on the focal plane can be computed using the basic radiometric equations presented earlier, or by using a raytracing (usually nonsequential) program. The classic method to deal with this problem is to install a cylindrical baffle around the hole in the primary mirror, which is also shown in Figure 3.12. Such a baffle may block (vignette) the in-field beam, and the decision about whether or not such vignetting can be tolerated must be made relative to the optical performance requirements (vignetting vs. stray light performance) of the system.

3.4.1.2 Specular Mirror Reflection

Like direct illumination of the detector, this path also is particular to reflective systems such as Cassegrain telescopes. An example of a specular mirror path is also shown in Figure 3.12: light from a source either inside or outside the FOV undergoes multiple bounces between the primary and secondary mirror, and goes to the detector. Typically, the only way to model this path is by using a nonsequential raytracing program. As with directed illumination of the detector, this path can sometimes also be blocked using a cylindrical baffle around the hole in the primary, and perhaps also by using a ring of black paint or anodization around the hole. Again, the decision about whether or not to block the path needs to be made relative to the requirements of the optical system.

3.4.1.3 Ghost Reflections

This is one of the most common types of stray light mechanisms, and is responsible for the prominent stray light artifacts shown in Figures 3.1 and 3.2. Any optical system with at least one refractive element (even if it is flat, such as a filter) will have ghost reflections, because there will always be some light reflected by the boundary between the air and the refractive material (even if the boundary is AR coated). Note that digital camera systems will also have a ghost reflection off of the detector, since all detectors reflect a small amount of light. An example of a path with ghost reflections is shown in Figure 3.13. Typically, two ghost reflections are necessary to couple light to the focal plane (i.e., the path must be second order), although ghost reflections can be paired with other mechanisms in the path, such as surface scatter. Since ghost reflections can occur from powered (i.e., curved) optical surfaces, they often come to a focus. A focused ghost path can be particularly problematic if the focus occurs close to the focal plane, since it will appear to be particularly bright. Ghost reflections usually occur from sources inside or just outside the nominal FOV, and can be modeled using sequential or nonsequential raytracing programs, though nonsequential programs are usually more flexible. Ghost reflections are typically mitigated through the use of AR coatings, or by changing the radius of curvature of one or more of the optics to prevent a ghost from focusing near the focal plane.

Ghost reflections also include total internal reflection (TIR) from one or more surfaces, which occurs when light strikes a lens surface from the inside and reflects off. This occurs primarily in lenses with surfaces that have small radii of curvature (and therefore usually high optical power). Light striking this surface at an AOI greater than the critical angle θ_c ($\theta_c = \sin^{-1}(1/n)$), where n is the index of the lens) will be totally reflected back into the lens, as from a mirror, as also shown in Figure 3.13. The AOIs of the rays in the primary optical path of a well-designed optical system are usually not large enough to undergo TIR; however, rays in stray light paths sometimes are, especially off

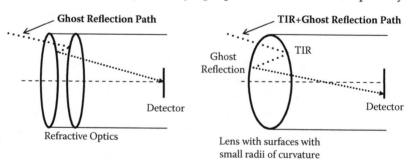

FIGURE 3.13
A ghost reflection path (left) and a TIR path from a lens surface with a small radius of curvature (right).

of the flange features found on the edges of some molded optics. Examples of such optics are shown in Figure 3.14, and an example of a TIR path that could occur from such an optic is shown in the Figure 3.15. A typical method used to mitigate this path is to roughen or apply black paint to the surfaces at which TIR occurs.

FIGURE 3.14
Molded optic lenses with flanges.

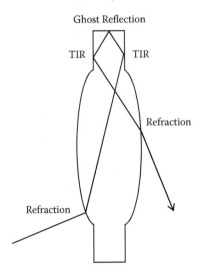

FIGURE 3.15
TIR path inside a molded optic with a flange.

3.4.1.4 *Specular Reflection from or Transmission through Mechanical Surfaces*

Smooth mechanical surfaces such as some types of plastic or polished metal can have strong specular reflections, and therefore it may be appropriate to model these surfaces as specular reflectors, though it's almost always better to use measured BRDF data if they're available (see Section 3.4.2). In many cases, however, these data aren't available, and in fact, all that may be known about the surface is that it visually appears to be specular (i.e., shiny), in which case this model may be appropriate. A classic example of a problematic specular reflection from a mechanical surface is shown in Figure 3.16. Light from outside the FOV strikes the inner diameter of the lens barrel, reflects, and hits the detector. Because the light is reflecting from a surface with a negative radius of curvature, it comes to a focus (though a poor one) and creates a caustic pattern. To model this path, it's usually necessary to use a nonsequential raytracing program. A typical solution to this problem is to use a baffle to block this path, as shown in Figure 3.16. Baffle design will be discussed in more detail in Section 3.5.6.2. If a baffle is not practical, the inner diameter of the barrel can be molded using a rougher mold or by using less reflective (i.e., black) plastic. If the barrel is metal, a black anodized treatment can be applied. For both plastic and metal barrels, a diffuse black paint can be applied.

In molded optics, flanges around a lens can be considered mechanical surfaces, and if they are not covered or roughened, they can result in specular paths (sometimes zero order) to the detector. An example of such a path is shown in Figure 3.3; light transmits through the uncovered flange to the detector. Roughening this surface or covering it with another surface (such as a baffle made of Mylar or other thin material) will mitigate this path.

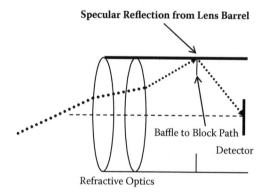

FIGURE 3.16
Specular reflection from a lens barrel.

3.4.2 Scattering Mechanisms

Scattering mechanisms redirect light in ways not predicted by Snell's laws of reflection and refraction. In order to accurately model these mechanisms, the BSDF of the mechanism must be estimated, and techniques for doing this will be discussed in this section. Before discussing these mechanisms, it is necessary to discuss BSDF and RMS surface roughness measurements.

3.4.2.1 BSDF and RMS Surface Roughness Measurements

The BSDF of any given surface is often a function of many parameters (such as manufacturing technique, cleanliness, composition of the surface, and the wavelength and polarization of incident light), and because of this, it is often difficult to develop highly accurate theoretical models of this scatter. For this reason, it is almost always more accurate to measure the BSDF of the surface than to use a theoretical model. In order to use the measured data, it is usually necessary to fit it to a function using an algorithm such as the damped least squares technique. However, BSDF measurement may not be practical in terms of the time and budget required, and may be difficult for inexperienced analysts. These competing factors must be weighed when deciding how to model the BSDF of a given surface. Fortunately, the models presented in this section are useful both empirically and theoretically.

BSDF measurement is done with a device called a scatterometer.[9] There are commercial companies that sell scatterometers and scatter measurement services.[10] For any given combination of surface and sensor type, these companies can recommend the measurements necessary to characterize it. In general, BSDF properties are not highly dispersive, and therefore are usually measured only at a single wavelength (the exception to this is when the surface is to be used in an extremely broadband sensor, such as sensor that has a visible and a long-wave infrared band). Most optical surfaces need to be measured at only one AOI, whereas mechanical surfaces usually need to be measured at at least three AOIs (usually 5°, 45°, and 75°). As BSDF measurements are made at discrete values of λ, θ_i, and θ_s, it is necessary to interpolate these data during analysis, and the models presented in this section are used for fitting this type of data.

As will be shown in the section on optical surface roughness, the RMS roughness of a surface can sometimes be used to determine the BSDF of a surface. This quantity is usually represented by the symbol σ or Rq, and is equal to

$$\sigma = \sqrt{\frac{1}{N}\sum_{i=1}^{N} z_i^2} \qquad (3.22)$$

where N is the number of points measured on the surface and z_i is the surface height of the ith point, as measured from the mean level. The quantity can be measured using devices such as a white light interferometer[11,12] or surface profilometer.[13] These devices are much more common than scatterometers, and therefore it is much more likely that the RMS roughness of a surface has been measured than its BSDF. A comprehensive discussion of the theory behind RMS surface roughness is beyond the scope of this book,[9] however, note that the correct spatial frequency bandwidth limits must be used when specifying RMS roughness.[14]

3.4.2.2 Scattering from Mechanical Surfaces

Mechanical surfaces scatter differently than optical surfaces, and therefore they are modeled differently. Mechanical surfaces are generally rougher, and therefore the scatter from them is much more diffuse. The depth of the surface roughness features of mechanical surfaces can result in more complicated variation in scatter vs. AOI, and the use of paints or dyes on the surface can result in more complicated variation in BSDF vs. wavelength. For these reasons, modeling of scatter from mechanical surfaces is usually done using empirical models. Three such models are the Lambertian model, the Harvey model, and the general polynomial model. Obviously, mechanical surfaces with high TIS such as a bare metal or white plastic surface are undesirable, and roughening and painting or anodizing the surface is recommended. For more demanding applications, the use of baffles may be required (see Section 3.5.6.2).

3.4.2.2.1 The Lambertian BSDF Model

Probably the most well-known BSDF model is the Lambertian model, which is equal to

$$\text{BSDF} = \frac{\text{TIS}}{\pi} \tag{3.23}$$

where TIS is the total integrated scatter of the surface, as defined above. A Lambertian surface is a surface whose scatter is totally diffuse, that is, a surface whose apparent brightness (or radiance) is constant for all values of scatter angle θ_s. Though no real-world surfaces are truly Lambertian, some are close, such as a smooth wall painted with matte paint or a white piece of printer paper. A special material called Spectralon[15] is engineered to be very Lambertian. This material is very fragile and is normally used only for sensor calibration in the laboratory. Because of its simplicity, it is tempting to use the Lambertian model as often as possible; after all, if the TIS of a surface is known or can be estimated, then the BSDF of the surface is known if it's Lambertian, and in the absence of measured data, sometimes this is the best model that can be developed. Most surfaces are not very Lambertian, and using it can result in large errors in the stray light estimate.

3.4.2.2.2 The Harvey Model

A very versatile and widely used model is the Harvey model,[16] whose functional form is given by

$$BSDF = b_0 \left\{ 1 + \left[\frac{\left| \sin\left(\theta_i\right) - \sin\left(\theta_s\right) \right|}{l} \right]^2 \right\}^{s/2} \tag{3.24}$$

where b_0, l, and s are the model coefficients. This model reduces to the Lambertian model for $s = 0$. A plot of a Harvey BSDF model vs. $\sin(\theta_s) - \sin(\theta_i)$ (also called $\beta - \beta_0$) is shown in Figure 3.17. The b_0 coefficient sets the maximum BSDF value, the l coefficient sets the roll-off angle (a feature noticeable in most BSDF distributions), and the s coefficient sets the slope of the BSDF distribution. This model is useful for fitting measured scatter data from smooth, fairly specular mechanical surfaces such as some types of metal and plastic surfaces. As will be shown later, this model is primarily used to model scattering from optical surface roughness.

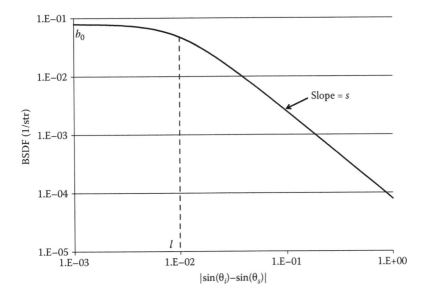

FIGURE 3.17
BSDF vs. $|\sin(\theta_i) - \sin(\theta_s)|$ of the Harvey model BSDF. This model corresponds to a molded optic surface with 40 Å RMS surface roughness.

3.4.2.2.3 The General Polynomial Model

This model is used to generate very accurate fits to scatter data measured from rough mechanical surfaces, and its function form is given by

$$\log(\text{BSDF}) = \sum_{k=0}^{n} \left[\sum_{i=0}^{m} \sum_{j=0}^{i} c_{ijk} \left(U^i W^j + U^j W^i \right) + \sum_{i=l'}^{l} c_{ik} \log\left(1 + d^i T\right) \right] \frac{V^k}{2} \quad (3.25)$$

where $U = \sin^2(\theta_s)$, $V = \sin(\theta_i)\sin(\theta_s)$, $W = \sin^2(\theta_i)$, $T = [\sin(\theta_s) - \sin(\theta_i)]^2$, and c_{ijk}, c_{ik}, and d are the model coefficients. A plot of a general polynomial model is shown in Figure 3.18. The advantage of this model is that it is very general, and because it is a polynomial of arbitrary order, it can be used to generate very accurate fits. The disadvantage of this model is that it can generate non-physical BSDF values for extrapolated values of θ_s and θ_i, especially if higher-order polynomial values are used. This can result in (among other things) values of TIS that are greater than unity for some values of θ_i, and therefore care must be taken when using this model to fit measured data. This model is most commonly used in the black paint BSDF models that are included with nonsequential raytracing programs such as FRED and ASAP. Coefficient values for a typical black paint scattering at visible wavelengths are given in Table 3.2, and the resulting BSDF is plotted in Figure 3.18. This plot shows

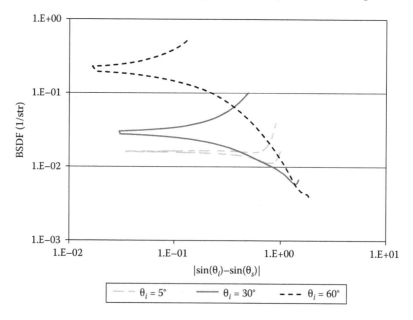

FIGURE 3.18
BSDF vs. $|\sin(\theta_i) - \sin(\theta_s)|$ of a typical black paint, modeled with the general polynomial BSDF model.

TABLE 3.2

General Polynomial BSDF Model Coefficients (c_{ijk}) for a Typical Black Paint at Visible Wavelengths (0.4–0.7 µm)

i	j	k 0	k 1
0	0	–1.802308	1.084313
1	0	–0.026716	–0.28679
1	1	0.7198765	–4.54178
2	0	–1.931007	4.445556
2	1	4.9618048	8.844519
2	2	–1.93257	–11.4638
3	0	2.7508931	–5.04243
3	1	–7.751616	7.44191
3	2	0.187639	–3.50296
3	3	4.1839095	4.560649

that the BSDF of the black paint increases with θ_i, which is typical of most black surface treatments, including black plastic and anodization.

3.4.2.3 Scattering from Optical Surface Roughness

No optical surface can be made to be perfectly smooth, and therefore all such surfaces will scatter light. Optical surfaces are always critical, and are very often illuminated by stray light sources as well, and therefore scatter stray light to the focal plane. The BSDF of an optically polished surface can be approximated using the Harvey model (Equation 3.24). This model can be used empirically to fit measured scatter data to an optical surface or theoretically to predict its scatter. The theoretical model requires that assumptions be made about the values of l and s, and then these values, along with the measured or estimated value of the RMS surface roughness σ of the surface, are used to compute b_0. A typical value of l is 0.01 rad, and s typically varies between –2 and –3 for most optical surfaces (the more negative the number, the more specular the surface). The roughness of an optical surface typically varies from 10 to 50 Angstroms (Å). Using these values, the refractive index of the surface n ($n = -1$ for mirrors), wavelength of incident light λ, and b_0 coefficient of the Harvey model can be computed using the equation

$$b_0 = \left[\frac{2\pi(n-1)\sigma}{\lambda}\right]^2 \left(\frac{1}{2\pi}\right)(s+2)l^s\left[\frac{1}{\left(l^2+1\right)^{(s+2)/2}-\left(l^2\right)^{(s+2)/2}}\right] \qquad (3.26)$$

for $s \neq -2$ and

$$b_0 = \left[\frac{2\pi(n-1)\sigma}{\lambda}\right]^2 \left(\frac{1}{\pi}\right)\left[\frac{1}{\ln\left(1+1/l^2\right)}\right] \tag{3.27}$$

for $s = -2$. Both of these equations contain the TIS of the rough surface, which is given by

$$TIS = \left[\frac{2\pi(n-1)\sigma}{\lambda}\right]^2 \tag{3.28}$$

b_0 and the TIS vary as $1/\lambda^2$, meaning that surface roughness scatter is a bigger problem in the visible ($\lambda \sim 0.5$ μm) than it is in the long-wave infrared ($\lambda \sim 10$ μm). The value of b_0 for a surface with $\sigma = 40$ Å, $\lambda = 0.45$ μm, $n = 1.5354$, $l = 0.01$ rad, and $s = -1.5$ (all typical values for molded optics) is $b_0 = 0.07906$ sr^-1 (TIS = 0.0894%). A plot of the BSDF of this surface is shown in Figure 3.17. The roughness of the molded optic is usually equal to the roughness of the mold used to make it, and therefore the optical surface roughness can be changed by changing the roughness of the mold. Scattering from optical surfaces that are AR coated is typically very similar to the scattering from the uncoated surface; however, surfaces that have bandpass filters or other coatings with many layers may scatter much more than predicted by the bare surface scatter alone. A thorough discussion of this phenomenon can be found in Elson.[17]

3.4.2.4 Scattering from Particulate Contamination

All real-world surfaces, whether they are mechanical or optical, have some amount of particulate contamination (dust), and this contamination increases the BSDF of the surface beyond the level predicted by its surface profile. If measured data are used to model the BSDF of a surface, and the measurement was performed on a surface whose cleanliness is representative of the cleanliness the surface will have in the final system, then the scattering from particulate contaminants is included in the BSDF data and no correction needs to be applied. However, if the measurement was performed on a surface that is much cleaner or dirtier than the surface in the system, or if the surface is modeled using a theoretical model such as the Harvey optical surface roughness model, then a theoretical model of the scattering from particulates must be used to obtain the correct BSDF of the surface. Such models have been developed using Mie scatter theory; however, they are complicated and discussion of them is beyond the scope of this book.[18] However, all of the nonsequential raytracing programs mentioned earlier include these models. A common input into these models is the Institute

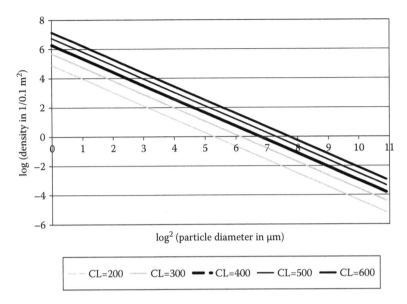

FIGURE 3.19
Particle density vs. particle diameter for IEST-1246C cleanliness distribution.

of Environmental Sciences and Technology (IEST) 1246C cleanliness level of the contaminated surface. This level is one method of quantifying the cleanliness of the surface; the higher the level, the dirtier the surface. The standard used to define it is derived from an older military standard (MIL STD 1246C), and defines the particle density vs. particle size distribution as a function of the cleanliness level. A plot of some typical particle size distributions for typical cleanliness levels is shown in Figure 3.19. A cleanliness level of 300 is a very clean surface and is usually achievable only by building the optical system in a clean room. A cleanliness level of 600 is a very dirty surface whose cleanliness level can be easily reduced by simple cleaning by hand. Optical surfaces in typical consumer digital cameras are at about cleanliness level 400, though the outermost surface may be higher. The BSDF (actually BTDF, or forward scatter) computed from the Mie scatter model in FRED is shown in Figure 3.20 for particulates on a plastic substrate, $\lambda = 0.45$ microns. The large peak in scatter at large scatter angles ($|\sin(\theta_i) - \sin(\theta_s)| \sim 1$) is due to high-angle scattering of large particles. Though the angular variation in BSDF vs. scatter angle from a contaminated surface is a function of wavelength (the longer the wavelength, the more specular the scatter), the TIS is not a strong function of wavelength, and is well approximated by the percent area coverage (PAC) of the particulates, which is given by[19]

$$TIS = PAC = \left(\frac{1}{100}\right) 10^{-7.245 + 0.926 \log_{10}^2(CL)} \qquad (3.29)$$

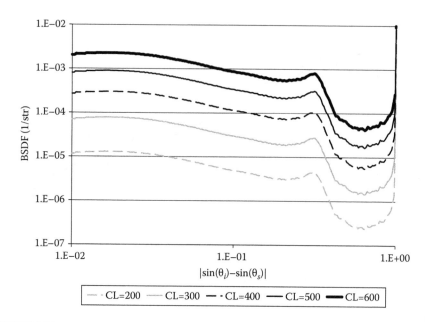

FIGURE 3.20
BSDF vs. $|\sin(\theta_i) - \sin(\theta_s)|$ of surfaces contaminated with IEST-1246C contamination distributions.

where CL is the cleanliness level. This equation is useful when comparing the amount of scatter from particulates to other scatter mechanisms (such as surface roughness). For instance, consider the 40 Å surface discussed in the section on optical surface roughness: the TIS of this surface due to roughness scatter is 0.089%, and is increased by 0.080% if it's contaminated at cleanliness level 400. This demonstrates that contamination scatter is a significant portion of the amount of scatter from the surface. A detailed discussion of contamination control and verification methods is beyond the scope of this book;[20] however, in general it is recommended that the optical system be assembled in a clean environment, and that the optical and mechanical surfaces are kept as clean as possible.

3.4.2.5 Scattering from Diffractive Optical Elements

As discussed in Chapter 1, many modern optical systems make use of diffractive optical elements (DOEs) to reduce the chromatic aberration of the system. DOEs are diffraction gratings etched or molded onto the surface of a lens or mirror, and as gratings they have a very high (near 100%) efficiency only at one wavelength, AOI, and diffraction order (usually the intended or "design" order, which is usually the +1 order). Since most DOEs are used across a band of wavelengths and AOIs, they will have less than 100% efficiency, and the light not directed in the design diffraction order will be scattered into other orders at other angles, as shown in Figure 3.21. In other words, an optical element with a DOE will scatter more than predicted by

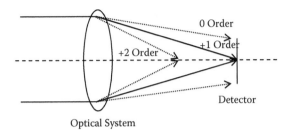

FIGURE 3.21
Raytrace of DOE diffraction orders. The +1 order is the design order.

surface roughness scatter and particulate contamination alone. The impact of a DOE on the stray light performance of a system is usually evaluated using a nonsequential raytracing program. Most programs require the input of a table of efficiency η as a function of the wavelength the DOE was designed to work at λ_0, the wavelength of incident light λ, and the diffraction order m (and sometimes the AOI θ_i). This efficiency equation is given by

$$\eta = \left\{ \frac{\sin\left\{\pi\left[\dfrac{\lambda_0}{\lambda\cos(\theta_i)} - m\right]\right\}}{\pi\left[\dfrac{\lambda_0}{\lambda\cos(\theta_i)} - m\right]} \right\}^2 \tag{3.30}$$

This equation assumes that the design order is +1. A plot of efficiency vs. wavelength for a DOE designed to work at $\lambda_0 = 0.55\ \mu m$ is shown in Figure 3.22. Notice that the +1 order reaches 100% efficiency only at the design wavelength, and that at all other wavelengths its efficiency is less and the efficiency of the other orders (and thus the amount of scatter) is nonzero. Also notice that the efficiency of the orders decreases as the order number deviates from the design order (+1). The angular deviation of the orders adjacent to the design order is small, and therefore the light from these orders comes to a focus very close to the focus of the design order. This leads to the classic stray light artifact from DOEs: small rings that appear around the images of light sources of narrow angular extent, such as streetlights at night.

In addition to the scatter from the diffraction orders, DOEs also scatter light from their transition regions, which are illustrated in Figure 3.23. These transition regions exist because the DOE grating surface cannot be made with infinitely sharp edges, and therefore the edges will have some small radius on them. The surface roughness of these radii is usually high, and therefore they will act as scatterers. Because they are so small, their scatter is difficult to measure, so a good starting point in assessing their impact is to assume they are Lambertian scattering surfaces and model them accordingly, either using first-order radiometry (assume the edges are flat annular regions in the

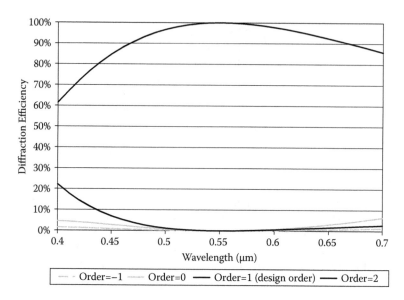

FIGURE 3.22
Diffraction efficiency vs. wavelength of a diffractive optical element (DOE).

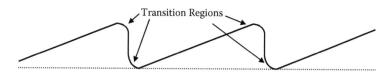

FIGURE 3.23
Transition regions of a DOE.

pupil) or a nonsequential raytracing program. As with particulate contaminants, the TIS of all of these regions can be estimated as their percent area covered (PAC), which is equal to the surface area of these regions divided by the area of the entire surface (note: this number is fractional, so it should always be less than or equal to 1). Obviously, the more transition regions there are, the more scatter there will be from the DOE. Thus, a DOE with a large number of diffraction zones will have higher scatter, a fact important to consider when designing the DOE.

3.4.2.6 Scattering from Haze and Scratch/Dig

Some particulate contaminants and air bubbles (also called "haze") will always be present inside a plastic or glass molded optical element, and these contaminants will scatter light. Plastic optics generally have more haze than glass, and the amount of haze increases with the thickness of the element. Haze is often characterized as a percentage, which is equal to the TIS (=PAC) of the haze particles. The haze values for typical plastic optics are given in

Table 4.1. In some cases, the TIS of the haze of a plastic optical element will be larger than the combined TIS of its surface roughness and contamination scatter, and therefore may be important to model.

This modeling can be done in a number of ways:

- Assume that haze scattering is Lambertian, which means that its BSDF is equal to (fractional haze)/p. This is a gross simplification and should be used only when no other options for modeling exist.
- Model the haze inclusions in a nonsequential raytracing program. These inclusions can be modeled as either spheres (for contaminants whose diameter is much greater than the wavelength of light) or, for contaminants whose diameter is less than or equal to the wavelength of light, using a volume scattering model such as the Henyey-Greenstein model. The PAC of the inclusions scattering model should be equal to the fractional haze.
- Scratches and digs can be modeled as appropriately-shaped sub-apertures of the optical surface (i.e., planes for scratches, disks for digs) with Lambertian scattering properties.
- Measure the BSDF of the optical element, which will include scattering from haze and scratches and digs.

3.4.2.7 Aperture Diffraction

Though technically not a scattering mechanism, aperture diffraction can result in stray light on the focal plane, and therefore should be considered in the design of the optical system. Figure 3.2 illustrates the effect of aperture diffraction; the star-shaped patterns emerging from the image of the street-lights are due to aperture diffraction from the camera aperture stop (iris). There are several texts that can provide a good understanding of diffraction,[22,23] and detailed evaluation of aperture diffraction can be conducted using coherent beam raytracing in either a sequential or nonsequential raytracing program. Note that the star-shaped pattern is just the two-dimensional Fourier transform of the aperture stop shape.

3.5 Stray Light Engineering Process

As discussed at the beginning of the chapter, this chapter is roughly divided into two sections: the first defined concepts and terminology necessary to perform stray light analysis, and the second, which begins here, discusses the application of these concepts in the design of an optical system. This design process is illustrated in the flowchart in Figure 3.24, and its basic form is similar to many other design processes: requirements for the system are

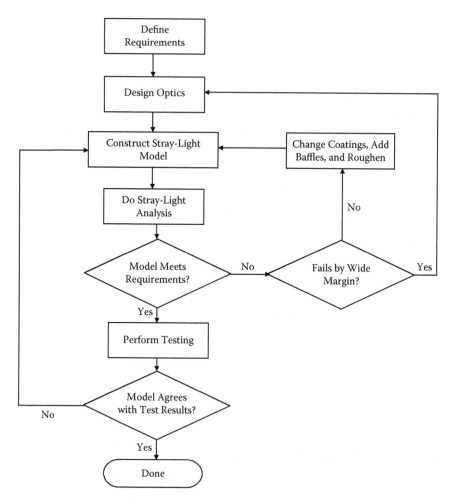

FIGURE 3.24
Stray light engineering process flowchart.

established, the initial system is designed, and its performance is evaluated relative to the requirements. If it does not meet the requirements, changes are made and its performance reevaluated, and this process is repeated until the requirements are met. This section of the chapter will discuss each of these process steps in detail, and the application of these steps will be illustrated by applying them to the molded optics cell phone camera design considered in Chapter 4, which is shown in Figure 3.25.

3.5.1 Define Stray Light Requirements

A stray light requirement defines what stray light performance is acceptable for a given optical system, and is determined by evaluating the purpose of

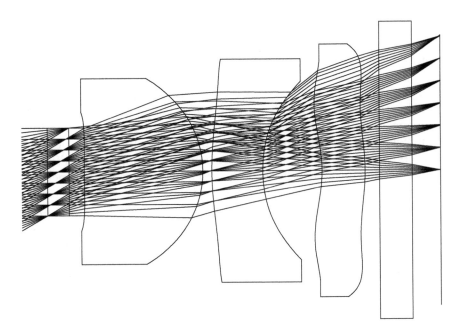

FIGURE 3.25
Molded optics cell phone camera lens from Chapter 4.

the optical system and the manner in which it is to be used. This is often one of the most difficult steps in the design process, because it requires an in-depth understanding of the purpose of the system, and because there is almost never a "perfect" set of requirements. Establishing requirements often involves trading off performance and ease of use with cost and complexity; the stricter the requirement is, the more difficult it will be to achieve, and thus the more complicated and expensive the system must be. A zero stray light requirement is not realistic: all optical surfaces have some roughness contamination and will scatter, and all refractive surfaces will produce ghost reflections, even if they are antireflection coated, and these mechanisms will produce stray light. It is impossible to list here all of the requirements that any conceivable molded optical system might require; this must be done by the optical system designers and by people familiar with the purpose of the system and the manner in which it is to be used. However, a few typical stray light requirements will be discussed here.

3.5.1.1 Maximum Allowed Image Plane Irradiance and Exclusion Angle

The maximum allowed image plane irradiance defines the maximum amount of light that can be on the image plane of an optical system from stray light sources. It can be defined for a single area of the image plane (such as the entire image plane), or for multiple areas of the image plane (the latter may be referred to as an image irradiance distribution requirement). For digital camera systems, the maximum allowed image plane irradiance is often set equal to the minimum detectable irradiance of the detector, or the irradiance that corresponds to a single grayscale bit. Obviously, if all of the stray light in the system is not detectable, then it is not a problem. This minimum detectable irradiance is usually a function of detector noise and is given by the detector manufacturer. It can be referred to by a number of different terms: minimum optical flux (i.e., watts) per pixel, noise equivalent irradiance (NEI), noise equivalent temperature difference (NEDT; used in infrared systems only), and others. For modern camera systems, these values are often very small, and thus it is rare for a system not to have some detectable stray light. Since the amount of stray light irradiance on the detector usually increases as the angle of the stray light source (i.e., sun, streetlights) to the center of the field of view (FOV) decreases, and since it is impossible to reduce all of this near-FOV stray light, the maximum allowed irradiance requirement is often accompanied by an exclusion angle requirement. The exclusion angle defines the minimum angle of the stray light source at which the maximum allowed image plane irradiance requirement is met. This geometry is illustrated in Figure 3.26. The exclusion angle is determined by setting the maximum allowed irradiance (either from the minimum detectable irradiance or from another method), doing a stray light analysis of the system, and then determining the source angle at which the irradiance requirement is first met. The exclusion angle requirement warns users of the system that sources near the FOV may result in a high level of stray light, and thus they may change the way they use the system in order to avoid this condition.

3.5.1.2 Veiling Glare Index

A typical stray light requirement used for visual camera systems (i.e., those intended for use with the eye) is the veiling glare index (VGI), which is defined as

$$VGI = \frac{\Phi_{out}}{\Phi_{out} + \Phi_{in}} \tag{3.31}$$

where Φ_{out} is the flux at the image plane due to uniform radiance outside the system FOV, and Φ_{in} is the flux at the image plane due to uniform radiance inside the FOV. This requirement is often used with a veiling glare test, which uses a large diffuser screen to illuminate the system from outside the FOV and is described in more detail in Chapter 7. The VGI quantifies the ability of the system to reject stray light coming from a source at any angle relative to the FOV, and tries to quantify the maximum amount of stray

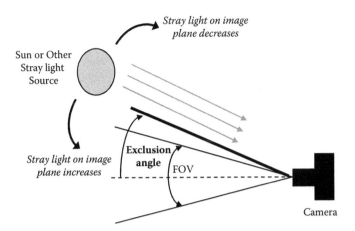

FIGURE 3.26
Exclusion angle geometry.

light the user of the system may observe (which is, of course, also a function of the brightness of the stray light source).

3.5.1.3 Inheritance of Stray Light Requirements from the Performance of Comparable Systems

This is not a stray light requirement in itself, but a method of defining requirements such as the maximum allowed image plane irradiance or veiling glare. This method often works well for consumer camera systems, since answering the question "What performance does the user expect?" is often easier than answering "What performance does the user need?" and assumes that the stray light performance of the comparable system is known or can be analyzed. Most modern consumer cameras, including digital single-lens reflex (SLR) and digital cell phone cameras, control stray light by using anti-reflection coatings on their optical surfaces and by roughening or blackening any mechanical surfaces that are near the main optical path. While these features reduce stray light, it is still detectable in these systems, especially in the lighting scenarios, such as the sun just outside the FOV and streetlights in the FOV at night, and especially when long exposures are used. However, this level of stray light control is acceptable for most consumer photography, and therefore the corresponding levels of stray light in these systems can be used to define the stray light requirement for new systems that will be used in a similar way. These types of systems usually result in a maximum stray light irradiance of about 3E14 photons/s-cm^2 (this value is on the high side, given that the focal plane irradiance from an average sunlit scene in the visible is about 7E14 photons/s-cm^2), and therefore this value is often a good starting point in establishing a requirement, and will be used in the evaluation of the example molded optical system considered here.

3.5.2 Design Optics

Once stray light requirements for the system have been established, the optical design of the system can begin. Though this step is concerned mainly with the optimization of image quality, it is important to know the stray light requirements of the system beforehand, since it may be necessary to incorporate features in the optical design to reduce stray light, especially if the stray light requirements are very strict. Some iteration may be required during this step, since the stray light performance of the optical system cannot be determined until the optical design is established. Two of the most commonly used features to control stray light in the optical design are intermediate field stops and cold stops.

3.5.2.1 Intermediate Field Stops

An intermediate field stop is an aperture in the optical system at an intermediate image, as shown in Figure 3.27 for a Keplerian telescope. The field stop is sized to be the same size and shape as the intermediate image (or usually slightly larger to account for manufacturing and alignment tolerances), and will block the light from any stray light source outside the FOV. This is one of the most effective ways to reduce stray light in an optical system, but it requires that the optical design include an intermediate image, and such designs are usually longer and more complex than ones without. Systems that are very intolerant to stray light, such as military, astronomical, or spaceborne systems, typically need field stops. More information about the use of field stops can be found in Smith.[6] Most consumer camera systems do not use field stops, since the reduction of size and cost is often more important than stray light performance in these systems. The example cell phone camera system considered in this chapter is such a system, and therefore no field stop will be used.

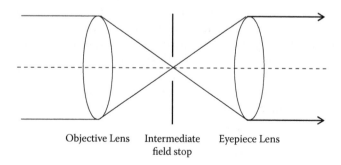

Objective Lens   Intermediate   Eyepiece Lens
field stop

FIGURE 3.27
A Keplerian telescope with an intermediate field stop.

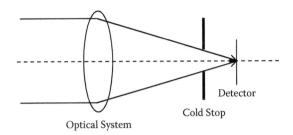

FIGURE 3.28
Optical system with a cold stop.

3.5.2.2 Cold Stops

A cold stop is an aperture used in an infrared optical system to limit the amount of warm housing that can be seen by the cold detector, and is usually part of the dewar assembly. If this stop is also the aperture stop of the system, it is called the cold stop; if not, it is called the coldshield. An example of a system with a cold stop is shown in Figure 3.28. Including a cold stop in the optical design is often necessary to control the amount of thermal background irradiance that reaches the detector, but requiring that the aperture stop be located close to the detector often results in a longer and more complicated optical design. Since the example cell phone camera considered here is not an infrared system, it does not need a cold stop.

3.5.3 Construct Stray Light Model

Once the optical design of the system is complete, the next step is to construct a stray light model. This involves modeling the optical and mechanical surfaces and assigning the appropriate specular and scattering properties to them. As mentioned earlier, Table 3.1 contains a list of surface types commonly found in optical systems and the appropriate specular and scatter models to use, which have all been discussed previously.

3.5.3.1 Analytic Model

Perhaps the simplest stray light model that can be constructed is the analytical model given in Equation 3.21. In this model, the BSDF of the optical system is approximated as the sum of the BSDFs of the illuminated optical surfaces. The use of this equation is a gross simplification of the stray light behavior of the system and neglects a number of important factors, such as the effect of the curvatures of the optical surfaces, shadowing of the optical surfaces by lens barrels and baffles, and the effect of specular stray light paths and scatter from mechanical surfaces. Due to these limitations, it is not recommended to use this model to compute the final estimate of the stray light performance of the system, and this model should only be used as the final estimate if it is not possible to construct a nonsequential raytracing model. In the example cell phone camera model, the BSDF of each optical surface can be modeled using the surface roughness Harvey model discussed in Section 3.4.2.3 for a 40 Å roughness ($l = 0.01$ rad, $s = -2.5$, $b_0 = 0.79062$ sr^−1). Since there are eight optical surfaces in this system, a worst-case assumption is to assume that they all are fully illuminated, and therefore the BSDF of the system $BSDF_{system}$ is

$$BSDF_{system} = 8b_0 \left\{ 1 + \left[\frac{\left| \sin(\theta_i) - \sin(\theta_s) \right|}{l} \right]^2 \right\}^{s/2} \tag{3.32}$$

3.5.3.2 Nonsequential Raytracing Model

As mentioned earlier, a more accurate way to construct a model is to use a nonsequential raytracing program. This program allows the optical design to be electronically imported from a sequential optical design program, and the mechanical geometry from a commercially available mechanical CAD program.[24,25] The details of importing geometry vary from program to program and will not be discussed here; please refer to the software vendor's documentation for these details. However, there are a number of best practices that should be followed when importing:

- After importing geometry from the sequential optical design program, always check an image quality metric in the nonsequential program to make sure it's the same, since the algorithms in these programs used to import geometry are not always bug-free. RMS spot size is an easy image quality metric to check in a nonsequential raytracing program.

- If the imported mechanical geometry contains representations of the optical surfaces (i.e., lens or mirror surfaces), they most likely will not raytrace correctly, and therefore should not be used. Most mechanical programs lack the ability to model these surfaces to the accuracy required, and therefore these surfaces should always be represented using geometry imported from the optical design program.

- Importing geometry from mechanical CAD programs can be problematic, and therefore it is best to do only when the geometry is very complicated and difficult to construct using surface types native to the nonsequential programs. The algorithms used to import mechanical geometry (which is usually defined in an IGES or STEP file) are complicated and can be error-prone. Other problems with imported mechanical geometry include the fact that it usually does not raytrace as quickly as native geometry, and it can be cumbersome to work within the nonsequential raytracing program because it may use an inconvenient coordinate system. For example, if the optical system being modeled uses a simple cylinder as a lens barrel, then this surface probably does not have to be imported from a CAD program, since most nonsequential programs can model cylinders natively.

Once the geometry is imported, specular reflectance and scattering models must be assigned to it. Again, the details of defining these models in the nonsequential programs are best explained in the user documentation for the programs. A recommended best practice is, whenever possible, always aim the scattering from any surface at the virtual image of the image plane, since this is the most efficient way to raytrace scattered rays. The user documentation for the program (or perhaps the short-course notes for stray light analysis using the program) will explain how to do this.

In addition to defining the geometry, sources of stray light must also be defined in the model. Typically these sources are external to the sensor, like the sun, or for infrared systems, they may be internal to the sensor (see Section 3.5.4.2.1 for more information about infrared systems). The simplest way of modeling the sun is as a point source at infinity, which illuminates the entrance aperture of the system as a collimated beam. Any stray light source of sufficiently narrow extent, such as a streetlight, can be modeled using a point source at infinity. Neglecting the angular extent of these stray light sources is often acceptable; however, if the system has a very small FOV or if high accuracy is required in the stray light simulation, then it may be necessary to model the extent of the source. For instance, the sun has an angular extent of 32 arc seconds or about 0.53°, and therefore in order for it to be out of the FOV, it must be at an angle of at least (FOV + 0.53)/2° from the center of the FOV. This is not true of a point source, which is out of the FOV at (FOV/2)° from the center of the FOV. If this difference is important, then the angular extent of the sun must be modeled. Instructions for doing this are given in the user's manual for the nonsequential raytracing program used.

The optical design for the example cell phone system was imported from ZEMAX into FRED, and the lens barrel was built up in FRED from native geometry. The dimensions of this native geometry were obtained from the mechanical drawing package for this system. In the absence of such information, it is often acceptable to use approximate mechanical geometry, especially if it is simple geometry consisting of just cylinders and planes. The resulting FRED model is shown in Figure 3.29. The L\lens at the entrance aperture will be referred to as element 1, or E1; the next element, E2, etc.

The following assumptions were made about the specular and scattering properties of the surfaces in this model:

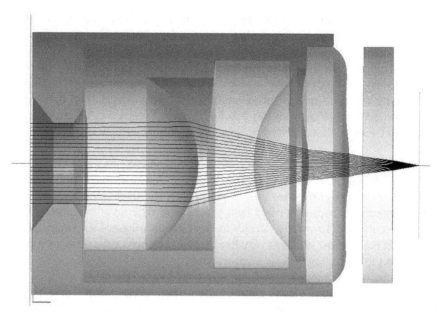

FIGURE 3.29
Stray light model of molded optics cell phone camera in FRED.

- All optical surfaces used the same Harvey roughness model used in the previous section (40 Å roughness), all are assumed to be at cleanliness level 400, and all are assumed to be uncoated surfaces.

- All mechanical surfaces are assumed to be shiny plastic with 50% specular reflectance.

As will be shown, the lack of AR coatings on the optics (which is not typical in final systems) and the high reflectivity of the mechanical structures will result in unacceptable stray light performance and will need to be changed.

3.5.4 Perform Stray Light Analysis

Once the model has been defined, the stray light analysis can proceed. This usually consists of defining the radiance of the stray light sources and their locations, and then determining the resulting image plane irradiance. Details of performing this analysis are different, depending on whether the analytic or raytracing model was used.

3.5.4.1 Analysis Using an Analytic Model

Use of the analytic model consists of determining values for all of the terms in Equation 3.21. In the cell phone camera analysis example, the following assumptions are made:

- The stray light source is the sun; therefore, the radiance of the stray light source L is equal to the blackbody integral of a $T = 5{,}900$ K blackbody over the visible waveband (0.4 to 0.7 μm), which is equal to 2.2316E21 photons/s-cm²-sr, and the projected solid angle of the stray light source Ω is equal to a right angle cone (Equation 3.10) that subtends 32 arc seconds, or $\Omega = 6.8052$E-5 sr.

- The scatter angle θ_s is roughly equal to zero (i.e., the scattered rays travel parallel to the optic axis).

- The transmittance of each optical surface is assumed to be 96% (uncoated); then transmittance of the system t is $(0.96)^8 = 0.8508$.

Given these assumptions, Equation 3.21 can be rewritten as the irradiance at the detector as a function of solar angle:

$$E(\theta_i) = L\cos(\theta_i)(\Omega)\text{BSDF}_{\text{system}}(\theta_i)\left[\frac{\pi}{4(F/\#)^2}\right]\tau \qquad (3.33)$$

This function is plotted in Figure 3.30, along with the maximum allowed image plane requirement established earlier in this section. This plot illustrates that the irradiance and the image plane fall off rapidly as the angle of the sun increases (this is typical), and that the maximum allowed irradiance requirement is met for all solar angles. However, the analytic model does not include all stray light mechanisms.

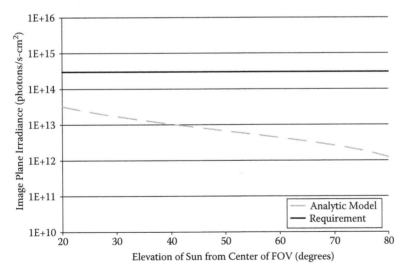

FIGURE 3.30

Stray light irradiance from the sun at the image plane vs. solar elevation angle, as predicted by the analytic model.

3.5.4.2 Analysis Using a Nonsequential Raytracing Model

Performing the analysis using the nonsequential raytracing model usually consists of defining the appropriate stray light sources in the model and performing the raytrace. As with other aspects of the nonsequential model, the details of performing this analysis vary from program to program, and therefore are best explained in their user documentation. However, a general rule to follow when using these programs is to perform both backward and forward raytracing.

3.5.4.2.1 Backward Raytracing

Though it may seem contradictory, the first raytrace that should be performed in any stray light analysis is a usually a backward raytrace from the image plane. This is done by defining an extended, Lambertian source at the image plane and tracing rays backward through the system to the entrance aperture and collecting them there. The main reason to do this is that usually the stray light source and many of the stray light paths (especially specular paths) have very narrow angular extents, and therefore finding all of these paths for the given stray light source can take a prohibitively long time. If the angular range in front of the sensor is not adequately sampled in the forward raytrace, then one or more stray light paths may not be identified. Backward raytracing greatly reduces the chance of this happening by densely sampling the angular space. In addition, backward raytracing is necessary for infrared systems to compute the irradiance at the detector due to self-emission of the sensor itself. This calculation, called a thermal background calculation, requires that the product of the projected solid angle (between an area on the detector and a piece of geometry in the sensor) and the path transmittance $\Omega\tau$ be computed. The irradiance on the detector E due to this self-emission is then

$$E = L\Omega\tau \tag{3.34}$$

where L is the blackbody radiance of the piece of sensor geometry. The $\Omega\tau$ product is computed from the backward raytrace using the equation

$$\Omega\tau = \frac{E_{\text{geometry}}}{L_{\text{BRS}}} = \frac{E_{\text{geometry}}}{\Phi_{\text{BRS}}/(\pi A_{\text{BRS}})} \tag{3.35}$$

where E_{geometry} is the irradiance on the piece of sensor geometry resulting from the backward raytrace and L_{BRS} is the radiance of the backward raytracing source, which is equal to the power of the source Φ_{BRS} divided by the area of the source A_{BRS} divided by π. Notice that Equations 3.34 and 3.35 are the same equation with the terms rearranged; this is a consequence of the

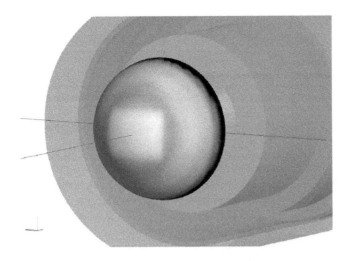

FIGURE 3.31

The detector FOV map showing the amount of light reaching the image plane as a function of the azimuth and elevation angles in front of the camera. The map is rendered on a hemisphere at the entrance aperture. The bright rectangle in the center corresponds to the nominal FOV, and everything outside this rectangle is the result of stray light. The grayscale is log 10 (intensity).

invariance of the A-omega product discussed in Section 3.3.1.8, and validates the use of backward raytracing.

A backward raytrace was performed for the example cell phone camera, and the intensity of the rays collected at the entrance aperture was plotted vs. angle in Figure 3.31. This plot is often called a detector FOV plot, and represents the throughput of the optical system as a function of source angles (azimuth and elevation) for the entire hemisphere in front of the sensor, including the throughput due to stray light paths. The bright rectangle in the center of this plot corresponds to the nominal FOV, and everything outside of it corresponds to stray light. Using the path analysis capabilities found in most nonsequential raytracing programs, it is possible to generate a list of stray light paths that end at the entrance aperture and thereby contribute to the detector FOV. These paths can be used to identify surfaces that are both illuminated and critical; any surface in these paths that has a stray light mechanism (ghost reflection, scatter, etc.) is both a critical and an illuminated surface. This list of paths was generated for the detector FOV plot shown, and the paths with the top ten flux levels are given in Table 3.3. The most significant path is the reflection from the lens barrel between the first and second lenses, as shown in Figure 3.32. In the detector FOV plot, this path results in the bright ring around the nominal FOV. Other significant paths include reflection from the cone (aperture stop) around the entrance aperture, as shown in Figure 3.33, and diffraction orders from the DOE. This

TABLE 3.3

Top Ten Stray Light Paths Identified in the Backward Raytrace

Percentage of Power at Entrance Aperture	Path Description
13.227%	Detector→E4→E3→E2→reflect from lens barrel between E1 and E2→E1
6.9842%	Detector→E4→E3→E2→reflect from lens barrel between E1 and E2→E1→reflect from cone at entrance aperture
3.143%	Detector→E4→E3→E2→E1→reflect from baffle cone at entrance aperture
1.8179%	Detector→E4→E3→E2→diffract at +2 order from DOE→E1
1.2487%	Detector→E4→E3→E2→diffract at 0 order from DOE→E1
0.6879%	Detector→E4→E3→E2→E1→reflect from baffle cone at entrance aperture→ghost reflect from E1
0.6770%	Detector→E4→E3→E2 →E1→scatter from roughness and contaminants on E1
0.4152%	Detector→E4→E3→E2→diffract at +3 order from DOE→E1
0.4061%	Detector→E4→E3→E2→diffract at +2 order from DOE→ reflect from lens barrel between E1 and E2→E1
0.1595%	Detector→E4→E3→E2→diffract at +4 order from DOE→E1

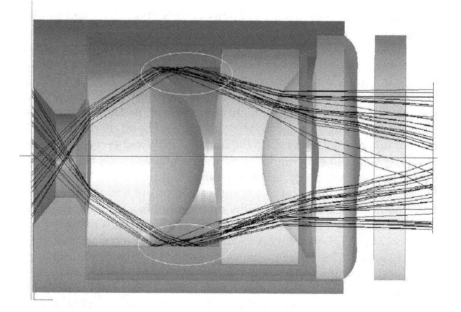

FIGURE 3.32
Reflection (circled) from the lens barrel between the E1 and E2 lenses. The barrel is modeled as a 50% reflective surface.

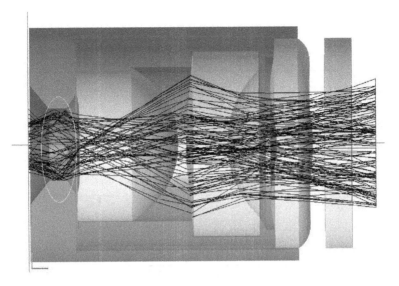

FIGURE 3.33
Reflection (circled) from the cylindrical baffle at the entrance aperture. The baffle is modeled as a 50% reflective surface.

backward raytrace identified hundreds of paths, including ghost reflection and scattering paths; however, none of these paths are as significant as the reflections from the lens barrel and DOE scatter paths. Though it will be necessary to perform the forward raytrace to compare the irradiance of these paths relative to the requirement, the results of the backward raytrace suggest that the lens barrel reflections are significant.

3.5.4.2.2 Forward Raytracing

Once the backward raytrace has been performed and the dominant stray light paths identified, the forward raytrace is performed. In a typical forward raytrace, the model of the stray light source (such as the sun) is set at a particular position, and rays are propagated from it through the system and allowed to scatter and ghost reflect to the image plane. Though the backward raytrace identified the surfaces that are both critical and illuminated, it is necessary to do the forward raytrace in order to determine which of these surfaces are illuminated for a particular stray light source at a particular location. If a surface is not on the list of both critical and illuminated surfaces generated by the backward raytrace, then no scattering from it needs to be modeled in the forward raytrace, since there is no way this scattered light will reach the image plane. The forward raytrace will also determine the spatial distribution of irradiance at the image plane for a particular stray light source at a particular position, something that the backward raytrace

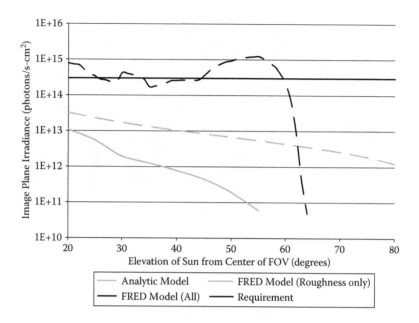

FIGURE 3.34

The average irradiance at the image plane due to stray light vs. solar elevation angle, as predicted by the analytic and FRED models.

does not compute. The information about the angular variation of stray light computed in the backward raytrace should be used when setting the angles of the stray light sources in the forward raytrace; those angular regions corresponding to high stray light flux should be well sampled in the forward raytrace. The resulting irradiance at the detector can then be compared to the requirement for a range of positions of the stray light source to determine if the system has adequate performance. This analysis was performed for the cell phone camera example for solar elevation angles varying from the edge of the FOV (about 20°) out to 80°. The resulting irradiance averaged over the entire image plane as a function of solar elevation angle is shown in Figure 3.34 for the system with just optical surface roughness scattering activated and with all stray light mechanisms activated, along with the results of the analytic model from the previous section. Notice that the FRED model with optical surface roughness scatter alone is similar to, but lower than, the analytic model results. The primary reason for this is that the analytic model assumes no shadowing of the optical elements closer to the detector by the lens barrel, whereas this phenomenon does occur in the FRED model and acts to reduce the stray light flux. Also notice that the FRED model cuts off the stray light at solar elevation angles greater than about 60°. This is due to shadowing by the conical baffle at the entrance aperture, and demonstrates the advantages of using such a baffle. The FRED model with all stray light mechanisms shows peaks in the stray light flux at solar elevation angles

FIGURE 3.35
The irradiance distributions at the image plane due to stray-light for solar elevation angles of 25° (a) and 55° (b). White corresponds to a focal plane irradiance of 4E16 photons/s-cm².

of about 25°, 35°, and about 55°. The peak at 25° is due to reflection from the cylindrical barrel at the entrance aperture (as shown in Figure 3.33), the peak at 35° is due to reflection from the lens barrel (as shown in Figure 3.32), and the peak at 55° is due to the reflection from the conical baffle at the entrance aperture. The irradiance distributions at the detector for solar elevations of 25° and 55° are shown in Figure 3.35; note that the irradiance over a subaperture of the image plane can be higher than the irradiance averaged over the entire plane. This is because the reflections from the lens barrel are specular. This analysis confirms that it will be necessary to mitigate the reflections from the lens barrel, as well as AR coat the optics.

3.5.5 Compare Performance to Requirements

At this point, if the estimated stray light performance meets the established requirements, then the analysis is done and the process can proceed to the testing phase. If not, then modifications must be made to the system and the performance reevaluated. As suggested in the process flowchart shown in Figure 3.24, it may be necessary at this point to consider the likelihood of meeting the established stray light requirement with the current design; if the estimated performance fails the requirement by a wide margin (for instance, if the current design has orders of magnitude more stray light irradiance than the established maximum allowed image plane irradiance), then changes to the optical design may be necessary, such as the addition of a field stop if one does not exist already. Of course, if such changes have already been made or are not possible, it may then be appropriate to consider whether or not the stray light requirement is achievable.

Figure 3.34 shows that, for the cell phone camera example, the requirement is not met for angles of the sun equal to about 20°, 35°, and 55°. Based on this information, a number of actions could be taken: the exclusion angle could be set to an angle greater than 55°, in which case the design is done and the process can proceed to the testing phase. However, it would not be practical to set the exclusion angle so large. Instead, blackening and baffles could be added to the system in an attempt to improve the stray light performance or, as suggested, the system could be redesigned with a field stop and the performance reevaluated. Given the need to make cell phone cameras compact, and given the relative ease of roughening the lens barrel surfaces, the best decision at this point appears to be roughen the surfaces.

3.5.6 Change Coatings, Add Baffles, and Blacken Surfaces

Changes such as the addition of AR coatings and roughening or blackening surfaces will increase the cost of the system, but not as much as redesigning the system with a field stop, and therefore they are commonly used. Stray light mechanisms and methods to mitigate these mechanisms are given in Table 3.4. These changes will be discussed here individually.

3.5.6.1 Change Coatings

Typically this step means adding AR coatings to uncoated surfaces. Though this increases the cost of the optics, its cost is small, and therefore AR coatings are used in nearly all optical systems (including molded optic systems) and are considered standard practice. The simplest way to add AR coatings to a stray light model is to increase the transmittance of the uncoated surfaces. For optics in the visible, this means changing the transmittance from 96% to 99%. A more accurate way of modeling AR coatings is to define their

TABLE 3.4

Stray Light Mechanisms

Mechanism	Severity	Modeling Method	Mitigation Strategy
Direct illumination	High	Raytracing	Optical/baffle design
Specular mirror path	High	Raytracing	Optical/baffle design
Ghost reflection path	Medium	Raytracing	Optical design, AR coatings
Optical surface roughness scattering	Medium	BSDF with first-order radiometry or raytracing	Material or optical polish selection
Diffraction	Medium	First-order radiometry or raytracing	Optical/baffle design
Contamination scattering	Medium	BSDF with first-order radiometry or raytracing	Cleanliness control
Scattering from diffractive optical element (DOE)	Medium	Raytracing	Limit the range of AOIs and wavelengths incident on the DOE
Scattering from mechanical structures	Medium	BSDF with first-order radiometry or raytracing	Optical/baffle design
Bulk scattering	Low	BSDF with first-order radiometry or raytracing	Material selection and processing

thin-film stack prescription in the nonsequential raytracing program, which allows the variation of the coating transmittance vs. AOI and wavelength to be accurately computed. These prescriptions can be difficult to obtain, since most coating vendors[26,27] consider these prescriptions to be proprietary. However, some textbooks[28,29] contain generic prescriptions, as well as information about coating design. For the cell phone camera example, a simple AR coating from Macleod (substrate + ¼ wave of MgF_2 + ½ wave of ZrO_2 + ¼ wave of CeF_2) was added to the lens surface. The transmittance of this AR coating vs. wavelength, as well as that of a typical uncoated molded optical surface, is shown in Figure 3.36.

3.5.6.2 Add Baffles

One method to deal with specular reflections and scattering from lens barrels and other pieces of mechanical geometry is to add baffles to it. The idea behind adding baffles is to block first-order stray light paths from reaching the image plane by preventing the geometry either from being illuminated or from being critical. This principle is demonstrated here in the design of optimal baffles for the cylindrical optical system shown in Figure 3.37a. Though this system does not contain any optical elements, the process used to baffle it is similar to the process used in any optical system. This system consists of just a circular entrance aperture, a cylinder, and a rectangular detector. The

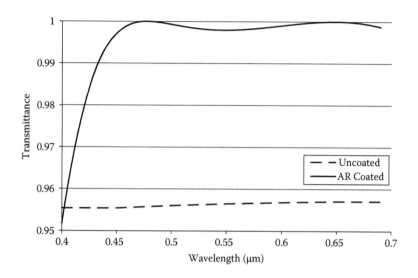

FIGURE 3.36
Transmittance of an uncoated and AR-coated molded optic surface.

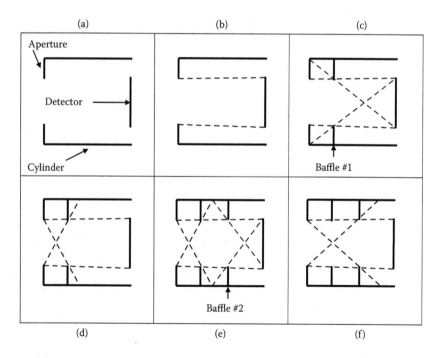

FIGURE 3.37
(a)–(f) (from upper left): Baffle design process.

longest radius of the detector (i.e., from the center to the corner) is shown in cross section in Figure 3.37a. The process for adding baffles is as follows:

1. First define a keep-out zone between the aperture and the detector, as shown in Figure 3.37b. This zone corresponds to the nominal optical path and cannot be blocked by baffles.
2. Draw a line from one edge of the detector to the opposite corner of the cylinder, as shown in Figure 3.37c. The intersection of this line and the keep-out zone line indicates the position of the first baffle, which prevents the portion of the cylinder between the entrance aperture and the baffle from being critical (i.e., from being seen by the detector).
3. Now draw a line from the edge of the entrance aperture to the opposite edge of the first baffle aperture, as shown in Figure 3.37d. This prevents the portion of the cylinder between the first baffle and the intersection point of this line from being illuminated.
4. Now draw a line from the edge of the detector to the intersection point on the cylinder of the previous line, as shown in Figure 3.37e. The intersection of this line and the keep-out zone line indicates the position of the second baffle, which prevents the illuminated portion of the cylinder between the first and the baffle from being critical (i.e., from being seen by the detector).
5. Go back to step 3 and repeat this process until no more baffles can be added, as shown in Figure 3.37f.

This baffle design prevents first-order specular or scatter paths from the inner diameter of the cylinder from reaching the detector. This process would be used exactly as described to design a coldshield in an infrared system with a cold stop, provided the coldshield did not contain any powered optics. In any system with powered optics, the process of placing baffles would be similar, except that backward raytracing from a Lambertian source at the detector would need to be used for steps 1, 2, and 4, and forward raytracing from a Lambertian source at the entrance aperture would be used for step 3. Be aware that all baffles have edges that will scatter, even if the edges are kept as sharp as possible (which is usually no smaller than a 0.005-inch radius). If too many baffles are used, then scattering from the edges may increase the amount of stray light rather than decrease it. In the cell phone camera example, baffles could be molded into the sides of the lens barrel; however, it will be shown that just roughening them is sufficient, and therefore baffles will not be used.

3.5.6.3 Blacken or Roughen Surfaces

Smooth, shiny mechanical geometry can cause stray light problems, as the initial analysis of the cell phone camera example shows. Blackening and roughening this geometry can reduce the magnitude of the stray light irradiance at the image plane. BSDF models for rough and black surfaces were discussed earlier in this chapter. The analysis of the cell phone camera indicated that reflections from the shiny lens barrel were the largest contributors to the stray light flux at the image plane, and therefore the performance of the system could be improved by roughening and blackening them. If the lens barrel is molded, then this may be done by using a rougher mold and by using plastic that has low reflectivity (i.e., black, for visible systems). It was assumed that the flat black paint model shown in Figure 3.18 accurately represents the scattering from rough, black molded plastic, and was applied to the lens barrel.

3.5.6.4 Improve Surface Roughness and Cleanliness

If the stray light analysis indicates that scattering from surface roughness and contaminants needs to be reduced, then this can be done for molded optics by decreasing the roughness of the mold and by improved control of particulates.[20]

3.5.6.5 Rerun Stray Light Analysis

As suggested by the process flowchart, the stray light performance of the system needs to be reevaluated after the improvements made and the performance compared to the requirements. The improvements made to the example cell phone camera were roughening the lens barrel and AR coating the lenses. The resulting performance computed from forward raytracing is shown in Figure 3.38; the requirement is now met at all angles. The largest improvement was made by roughening the lens barrel, though it is still the most significant contributor to stray light at the focal plane. The focal plane irradiance distributions now contain no bright artifacts as before, as shown in Figure 3.39, which was made by combining the image of a typical scene with the stray light focal plane irradiance distributions computed in the forward raytrace. This operation can be performed with image processing software,[30] taking care to scale the brightness of each image by its peak irradiance value.

3.5.7 Testing

Once the analysis of the system indicates that it will meet its stray light requirements, a prototype system should be built and tested. Stray light testing is highly recommended, as it is often difficult to know the accuracy of the BSDF models used in analysis; even if measured data are used, it may be difficult to know how similar the surface used in the as-built system is to the

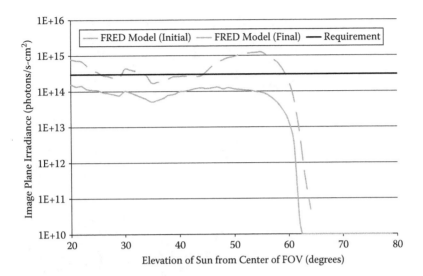

FIGURE 3.38
The average irradiance at the image plane due to stray light vs. solar elevation angle, as predicted by the FRED models, before and after lens barrel roughening and AR coating application.

sample used in the BSDF measurement. Methods for stray light testing will be discussed in Chapter 7.

3.6 Summary

The stray light performance of a molded optic system can be designed to meet the needs of the end user by using the stray light engineering process flowchart shown in Figure 3.24 to guide the design phase. In order to apply this process effectively, it is necessary to understand basic radiometry, the mechanisms by which stray light can reach the focal plane, and the methods by which these mechanisms can be modeled. Before designing the system, it is important to establish its stray light requirements so that the appropriate features can be added to the design, since it is often difficult or impossible to introduce them once the system is built. There are many commercially available nonsequential raytracing programs that can be used to accurately predict the stray light performance of the system and evaluate the effect of stray light control mechanisms, though it is important to also perform stray light testing of the system to verify that the requirements are met. There exist a variety of stray light control mechanisms that can improve the stray light performance of a system. Stray light testing of the as-built system is recommended.

FIGURE 3.39
Typical scene image irradiance distributions combined with the stray light irradiance distributions computed in the forward raytrace for the system before (top) and after (bottom) lens barrel roughening. The sun was assumed to be at a 55° elevation angle. The magnitude of stray light irradiance distribution for the before case (which is also shown in Figure 3.35b) had to be scaled down in this image to avoid saturating the pixels.

References

1. Optical Research Associates. CODE V software. http://www.opticalres.com.
2. ZEMAX Development Corp. ZEMAX software. http://www.zemax.com.
3. Photon Engineering LLC. FRED software. http://www.photonengr.com.
4. Breault Research Organization. ASAP software. http://www.breault.com.
5. Lambda Research Corp. TracePro software. http://www.lambdares.com.
6. Smith, W. J. 2008. *Modern optical engineering*. 4th ed. New York: McGraw-Hill.
7. Palmer, J. M., and B. Grant. 2009. *The art of radiometry*. Bellingham, WA: SPIE Press.
8. Wolfe, W. L. 1998. *Introduction to radiometry*. Bellingham, WA: SPIE Press.
9. Stover, J. C. 1995. *Optical scattering, measurement and analysis*. Bellingham, WA: SPIE Press.
10. Schmitt Measurement Systems. www.schmitt-ind.com.
11. Veeco Instruments. www.veeco.com.
12. Zygo Corp. www.zygo.com.
13. Taylor-Hobson Ltd. www.taylor-hobson.com.
14. Dittman, M. G., F. Grochocki, and K. Youngworth. 2006. No such thing as σ: Flowdown and measurement of surface roughness requirements. *Proc. SPIE* 6291.
15. LabSphere, Inc. http://www.labsphere.com.
16. Harvey, J. E. 1976. Light-scattering characteristics of optical surfaces. PhD dissertation, University of Arizona.
17. Elson, J. M. 1995. Multi-layer coated optics: Guided-wave coupling and scattering by means of interface random roughness. *JOSA A* 12(4):729–38.
18. Spyak, P. R., and W. L. Wolfe. 1992. Scatter from particulate-contaminated mirrors. Part 1. Theory and experiment for polystyrene sphere and λ = 0.6328 μm. *Optical Eng* 31(8):1746–56.
19. Ma, P. T., M. C. Fong, and A. L. Lee. 1989. Surface particle obscuration and BRDF predictions. *Proc. SPIE* 1165:381–91.
20. Tribble, A. C. 2000. *Fundamentals of contamination control*. Bellingham, WA: SPIE Press.
21. Henyey, L., and J. Greenstein 1941. Diffuse radiation in the galaxy. *Astrophys J* 93:70–83.
22. Gaskill, J. D. 1978. *Linear systems, Fourier transforms, and optics*. Boston: Wiley.
23. Goodman, J. W. 2005. *Introduction to Fourier optics*. 4th ed. Greenwood Village, NY: Roberts and Company.
24. Parametric Technology Corp. Pro Engineer software. http://www.ptc.com.
25. Dassault Systémes. SolidWorks software. http://www.solidworks.com.
26. Barr Associates. www.barrassociates.com.
27. JDS Uniphase Corp. www.jdsu.com.
28. Macleod, H. A. 2010. *Thin-film optical filters*. 4th ed. Boca Raton, FL: Taylor & Francis Group.
29. Baumeister, P. W. 2004. *Optical coating technology*. Bellingham, WA: SPIE Press.
30. Adobe Systems, Inc. PhotoShop software. http://www.adobe.com.

4

Molded Plastic Optics

Michael Schaub

CONTENTS

4.1 Introduction

Molded plastic optics are currently utilized in a wide variety of fields. Once relegated by both quality and perception to low-end products such as toys, their use has greatly expanded in the past few decades. The increase in the use, quality, and capability of molded plastic optics has been driven by improvements in both plastic optic materials and the equipment used to produce molded optics from them. Additionally, the growing reliance on optical technologies has provided markets and applications that did not previously exist for molded plastic optics.

In this chapter we discuss the properties of plastic optic materials, methods of producing molded plastic optics, and design considerations for their use. We begin by reviewing the commonly used materials, comparing their properties with one another. Next, the discussion is devoted to the manufacture of molded plastic optics, with emphasis on standard injection molding. Design considerations are then covered, along with a brief review of a cell phone camera lens. We also discuss aspects of prototyping and moving a design to production. Finally, we consider the future of molded plastic optics.

4.2 Materials

Compared to optical glasses, there are relatively few moldable optical plastics, as can be seen by examining the glass map shown in Figure 1.1. Even with this limited choice, there is a sufficient range of materials to enable designs that will perform adequately for many applications. An n-V diagram displaying only plastic optic materials is shown in Figure 4.1. As with optical glasses, plastic optic materials are divided into two categories, crowns and flints, based on their dispersion. The commonly used crown materials are polymethyl methacrylate (acrylic or PMMA), cyclic olefin polymers (COPs), and cyclic olefin copolymers (COCs), while the common flint materials are polycarbonate (PC), polystyrene (PS), and NAS, which is a copolymer of acrylic and styrene. In addition to these materials, there are a number of less commonly used materials, such as the crown materials optical polyester (O-PET) and polymethylpentene (PMP), and the flints polyetherimide (PEI) and polyethersulfone (PES). Some properties of these materials can be seen in Table 4.1 and are discussed below with reference to it.

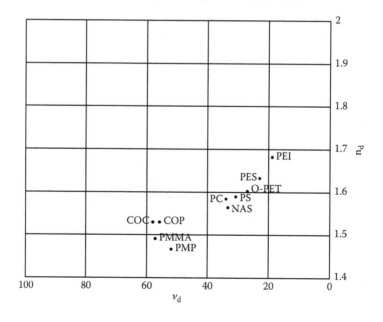

FIGURE 4.1
Glass map for common molded plastic optic materials.

TABLE 4.1

Properties of Common Molded Plastic Optic Materials

Material	PMMA	COP	COC	PC	PS	NAS	PMP	O-PET	PEI	PES
Glass code	491.572	530.558	530.580	585.340	590.308	564.334	467.520	602.270	682.189	633.230
Service temp. (°C)	92	130	130	124	82	80			170	160
dn/dt (e-6/C)	−85	−80	−101	−130	−120	−140		−130		
Vis. trans. (%)	92	91	91	88	87	90	90	86	50	70
Haze (%)	1–2	1–2	1–2	2	3	3	5			
Water absorp. (%)	0.3	<0.01	<0.01	0.15	0.2	0.15		0.15	0.25	0.22
Birefringence	4	1	1	7	10	5		1		
Spec. gravity	1.18	1.01	1.02	1.25	1.05	1.09		1.22	1.27	1.24
Color	Clear	Clear	Clear	Clear	Clear	Clear	Slight yellow	Clear	Yellow	Yellow

The table is arranged with the common crown materials in the columns on the left, the common flint materials to their right, and the additional materials further right. The top row displays the material type, while the second row displays the glass code for the material. As stated previously, the glass code is a six-digit number providing the refractive index (n_d) and the Abbe number V_d of the material. We can see that the common crown materials all have Abbe numbers approximately equal to 50, while the common flints have Abbe numbers in the lower 30s. For the less commonly used materials, PMP is a crown with an Abbe number of 52, while O-PET is a flint with an Abbe number of 27. PEI and PES are both flints with low Abbe numbers near 20.

The service temperature of a material is the sustained temperature that the material can be subjected to without significant degradation in its performance. There is no strict definition of service temperature, which should not be confused with standardized material tests such as heat deflection temperature. A review of various vendor websites will show different temperatures for the same material types. In some cases, this may be due to the vendor's preference for a specific grade of the material from a particular manufacturer. Often, it is based on actual project experiences of the company. We can see from the table that there is a wide range of service temperatures for plastic optic materials. For the common crowns, the COPs and COCs have higher service temperatures than acrylic. For the common flints, polycarbonate has a significantly higher service temperature than polystyrene or NAS. Polycarbonate is often paired with either a COC or COP material in designs that must withstand a higher temperature environment. For the less common materials, PEI and PES both have high service temperatures. O-PET has a melt transition temperature about 10°C higher than PMMA, implying that it should have about a 10°C higher service temperature.

The change in refractive index with temperature (dn/dt) is considerably higher for molded plastic materials than for optical glasses, including moldable glasses. From the table we can see that the values of dn/dt of plastic optic materials go from −80 to −140 ppm/°C. For comparison, the dn/dt of Ohara L-BAL42, a common molded optic glass, is 3.3 ppm/°C. In contrast with most glasses, the sign of dn/dt for optical plastics is negative, meaning that the refractive index of the plastic materials decreases with increasing temperature and increases with decreasing temperatures. As previously stated, the focal length of an optical element is related to the refractive index of the material it is made from. Thus, as the temperature changes the focal length of the plastic optic element will also change, possibly resulting in a change in focus.

The transmission of the optical plastics is quite good over the visible spectrum. The exceptions to this, for the materials in the table, are PMP, PEI, and PES. PMP can have a slight yellowish color, while PEI and PES appear a deep yellow or amber. This reduces their transmission in the shorter wavelengths, as can be seen in Figure 4.2, which is an internal transmission plot for PES. Transmission curves for some of the commonly used materials are shown in

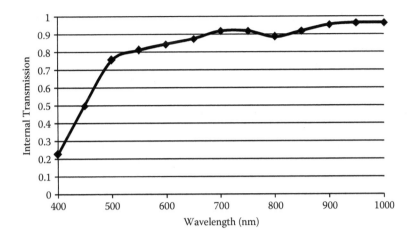

FIGURE 4.2
Internal spectral transmission of PES.

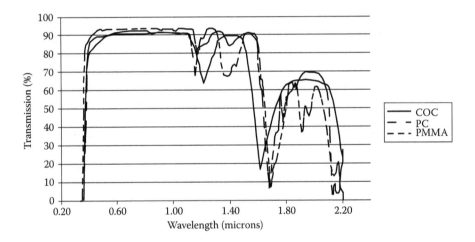

FIGURE 4.3
Spectral transmission curves for several common plastic optic materials.

Figure 4.3. For the common materials, as we approach the short end of the visible spectrum, the ultraviolet (UV) region, the transmission of the materials drops sharply. There is some variation in the cutoff wavelengths with material, providing options for systems that need to operate in the shorter-wavelength region. Different grades of a material may also have different UV cutoffs, with some grades specifically designed for short-wavelength transmission. At the other end of the spectrum, the near-infrared and short-wave infrared (SWIR), the materials have a number of absorption bands, resulting in reduced transmission. These absorption bands are associated with the chemical structure of the materials and can be seen across the material types. These absorption bands make it more difficult to use plastic optics for broadband imaging across the SWIR region.

It is possible to use plastics for wavelengths longer than the visible for certain narrower-wavelength band regions, such as are seen in laser systems. For instance, plastic optics may be used with Nd:YAG lasers at 1.06 microns, or at the 1,550 nm range, as is done in telecommunications. Care should be exercised when considering the use of plastic optics near the edge of an absorption band, as small shifts in the absorption of the material or the source wavelength can produce a dramatic drop in system transmission. Use of long-wave infrared plastic optics is possible using thin material sections, such as Fresnel or diffractive lenses.[1] To date, there is very limited use of plastic optics in the standard mid-wave infrared (MWIR, 3 to 5 microns) or long-wave infrared (LWIR, 8 to 12 microns) regions due to a lack of high-transmission materials.

In general, molded plastic optic materials have slightly more scatter than glass. There are standard material haze tests that can be performed to evaluate plastic scatter, such as ASTM-D1003. Haze will vary with the thickness of a sample and is usually not a concern in plastic optical systems unless the components are very thick and the material used is on the high end of the haze scale. The values for haze shown should be used as a comparison between the various materials. Most of the optical plastics have relatively low scatter, with the exception of PMP, PEI, and PES, which have slightly more haze. Viewing a sample molded plastic lens should reduce concerns about haze. If scatter is a critical issue, the use of plastic molded optics should be reviewed and evaluated against other alternatives, such as glass.

Plastic materials are known to absorb water. The amount of absorption is determined using standardized tests, where samples are conditioned, weighed, soaked in water for specific time periods, and then weighed again. The amount of water absorption can be calculated from these measurements. The newer materials, COC and COP, were designed to have low water absorption. The effect of water absorption is similar to that of a temperature increase, in that the part will swell slightly. For systems that will operate in humid or watery conditions, prototype systems should be built and tested.

Plastic optics are more prone to birefringence, which is the variation in refractive index with polarization state, than glass materials. This is partly due to the molding process, which can induce stress into the parts and create birefringence. The numbers shown in the chart are meant to be used as a relative comparison between the materials. The newer materials, COC, COP, and O-PET, have been specifically designed to have less birefringence than polycarbonate or polystyrene. While molding can induce birefringence, adjustment of the molding process can often have a large effect on the final birefringence state of the part. If birefringence is a concern, the molder should examine the parts using a crossed polarizer arrangement to assist in optimizing the molding process.

An important reason for the use of plastic optics, other than cost, is their reduced weight relative to glass. All of the materials in the table have specific gravities between 1 and 1.27. In comparison, the optical glass Ohara L-BAL42 has a specific gravity of 3.05. This difference in specific gravity can make molded plastic optics significantly lighter than glass optics. This weight advantage is often used in systems such as head-mounted displays, where reducing the amount of weight that needs to be carried on the head is a system design driver.

All of the material data in the table should be taken as guidance for the selection of materials for a project, but should not be used as stringent, unchanging design data. The values have been taken from various sites and will vary slightly with material manufacturer and grade. The designer is advised to verify the values he or she is using with the material manufacturer and molder before finalizing a design.

Once material types have been selected and a design developed, it is important that the materials be properly specified. Simply calling out the material as "acrylic" is not a sufficiently detailed specification. In general, the material type (e.g., acrylic), a material manufacturer, and a specific material grade should be provided. The reason for this is the large number of material grades for a specific material type. For instance, there are many different grades of acrylic, with the various grades containing different additives. In some cases, one grade may have a UV stabilizer present, while another grade may not. Some grades may contain mold release agents, which are not often used in molded plastic optics. Besides additives, different material grades can have different chemical or optical properties, such as UV cutoff.

In addition to calling out a specific material, the designer should also state what material substitution, if any, is allowed. For example, a particular molder may not have the specific acrylic material that was designated, because it prefers to use a similar acrylic from another material manufacturer. This could be because it has a great deal of molding experience with one material and not with the other. As such, the molder may wish to substitute its preferred material for the designated material. Depending on the wording on the drawing or purchase order, this may or may not be allowed. It is a good practice to include a note on the drawing stating that substitution for the designated material may take place only after written approval has been received from the customer. In this way, the designer will have the chance to evaluate the molder's recommended alternate material and determine if it is acceptable.

In addition to specifying a particular material type, manufacturer, and grade, the designer should also state that the material must be in an "unused" or "virgin" state. In some cases, excess material from the molding process or parts that do not meet specifications is ground up and fed back into the molding machine. This material, known as regrind, should generally not be used in molded plastic optics, as the properties of the material can be affected by repeatedly going through the molding process. Most optical parts molders are aware of this and do not use regrind. However, it is better to err on the side of caution and explicitly state on the drawing that the use of regrind is not permitted.

4.3 Manufacturing

The primary method of molding plastic optics is injection molding, where molten plastic is injected into a mold and allowed to harden, forming the desired component. The production of an injection molded plastic optic relies upon the use of injection molding machines and injection molds, as well as an optimized molding process.

4.3.1 Injection Molding Machines

A schematic of an injection molding machine is shown in Figure 4.4. This schematic shows a standard, horizontally oriented injection molding machine, though vertically oriented machines exist as well. The injection molding machine consists of two main mechanisms, the clamping mechanism and the injection mechanism. The clamping mechanism is used to close, hold, and open the mold, while the injection mechanism is used to inject plastic into the mold.

Injection molding machines are sometimes referred to as presses, in reference to the clamping mechanism. While standard injection machines do not directly press on the material in the mold, except at the injection port, the clamping mechanism must be able to exert a high pressure on the mold itself. Injection of plastic into the mold can generate pressures of thousands of kilos per square centimeter, requiring the clamping mechanism to have an even higher capacity, in order to keep the mold shut. Molding machines are rated by their maximum clamping force, usually expressed in tonnes.

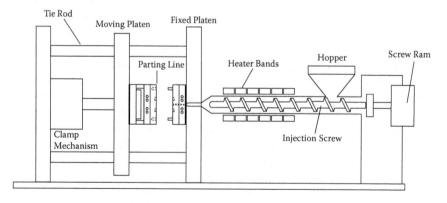

FIGURE 4.4
Schematic drawing of an injection molding machine.

Generally, the higher the clamping force is, the larger the size of the molding machine. Larger machines can be used to hold large molds, which allows more parts and bigger part sizes to be produced. Injection molding machines range in clamp tonnage from about 5 to 2,500 tonnes. Those used specifically for molded optics typically run in the range from about 10 to 50 tonnes. In addition to keeping the mold closed during injection, the clamping mechanism also moves to open and close the mold, and helps with the removal of the parts from the mold, as will be discussed later.

The clamping mechanism consists of two large metal plates, a means to provide the clamping force, and a cylinder/toggle to apply the force. The metal plates are referred to as platens. One of the platens has a fixed position in the molding machine and is referred to as the fixed platen. The other platen is attached to the clamping cylinder and moves back and forth to open and close the mold. Not surprisingly, this platen is referred to as the moving platen. Early molding machines used hydraulics to provide the clamping force. Today, many machines have switched to electrical servo mechanisms, leading to them sometimes being called all-electric machines. While hydraulic machines are still in use, many plastic optic molders have moved to electric machines, feeling that they provide superior control and repeatability. Others use a hybrid machine, which combines the desired aspects of the hydraulic and electric machines.

The injection mechanism of the molding machine consists of a hopper to hold the plastic material and the injection screw and barrel. The barrel is surrounded by a heating mechanism, such as heater bands, which aids in melting the plastic and bringing it to the desired temperature for injection. The injection screw is aptly named, as it simply looks like a large screw. For molding plastic optics, the screw is approximately a meter long, with its dimensions matched to the machine and mold size being used. The amount of material that a screw can inject into the mold at a single time is known as the shot size. Injection screws are made from a variety of materials, with tool steel often used. The injection screw selected is based on the material that is being molded, as the screw material and thread design can be optimized for certain material types and molding conditions. Using the wrong screw can result in a material interaction with the plastic being molded, or in poor filling of the mold. Most molding facilities have a variety of screw types and sizes, along with a range of injection molding machine sizes. This allows them to optimize the molding setup for a given material and mold. In addition to injecting the plastic into the mold, the injection mechanism also seals the mold from the injection side, maintaining pressure on the plastic that it injected.

4.3.2 Injection Molds

Injection molds are a critical component in the production of high-quality molded plastic optics. Molds, also referred to as tools, are typically composed

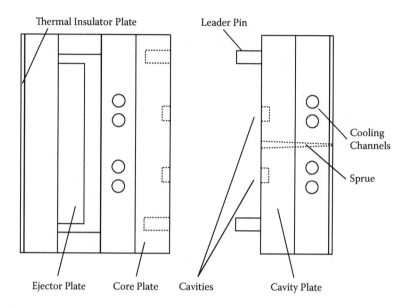

FIGURE 4.5
Schematic drawing of an injection mold.

of a series of plates containing various inserted components. The mold is divided into two halves, one of which is attached to the fixed platen, while the other is attached to the moving platen. In keeping with the platen naming convention, the two halves are referred to as the fixed and moving mold halves. Other terminology is also used, such as the hot and cold halves, the A and B halves, or the cavity and core halves. The surface where the two halves of the mold separate (and join) is referred to as the parting line. A schematic view of a mold is shown in Figure 4.5, and the two halves of an actual mold are shown in Figures 4.6 and 4.7. Referring to these figures, we discuss the function of the various plates in the mold. On its outside, each mold half has a thermal insulator plate. This plate insulates the mold, which is heated to an elevated temperature, from the platens, which are at room temperature. Interior to the insulation plate, each mold half has an attachment plate. These plates are used to connect the mold to the platens. On the fixed half of the mold, on the right in the diagram, the attachment plate is also labeled as a cooling plate. Referring to Figure 4.6, we can see a series of hoses connected to this plate. The hoses are used to flow oil through channels in the cooling plate. The oil will carry away heat from the molten plastic, helping it to cool and harden within the mold. The oil temperature is normally above room temperature, so it also serves to heat the mold. In some molds, internal electrical heaters are also used to control the mold's temperature. Mold temperature is one of several parameters that may be adjusted to optimize the mold process, as will be discussed later. The oil temperature is regulated by

FIGURE 4.6
Fixed half of an injection mold.

a thermal controller. Moving further inward on the fixed mold half, we reach the cavity plate. The cavity plate contains inserts that will form the surfaces of the molded optic component. In the case of the mold shown in Figures 4.6 and 4.7, it contains inserts to form a seven-lens array with a supporting structure. It can be seen that the mold contains four such sets of inserts, making this a four-cavity mold. Thus, each time the mold is put through a molding cycle, four copies of the part will be produced. It is common for molds to produce multiple parts, shortening the time required to produce a given number of parts, and generally reducing their cost. Looking at Figure 4.6, we see a central hole and a series of channels cut into the face of the cavity plate. During the injection process, the plastic enters the mold (the parting line region) through the hole in the cavity plate. Each plate exterior to the cavity plate has a similar hole, with the tunnel formed by them tapering down as it moves from the cavity plate face to the injection plate. This tunnel is known as the sprue. As the plastic is injected, it passes through the channels on the face of the cavity plate, which are known as runners, to the regions where the part-forming inserts are located. These regions are known as the cavities of the mold. The runners usually have a smooth, rounded profile, as sharp corners make it more likely that the runner (which will be filled with plastic) will stick during its removal. Connecting the runner to the cavity is a small opening known as the gate. The gate, along with the injection speed and pressure, determines how the plastic material flows into the cavity region. The gate's size and form are usually highly controlled in order to optimize

FIGURE 4.7
Moving half of an injection mold.

the material flow. For most molded plastic optics there is only one gate per part, as opposed to conventional molding, where multiple gates and runners per part are sometimes used.

Switching to the moving side of the mold, on the left in Figure 4.5, after the thermal insulation and attachment plate is a deep plate with a large cutout and another, smaller, internal plate. This is followed by a cooling plate and the core plate, which is akin to the cavity plate on the other side of the mold. The large plate with the cutout is known as the ejector box, while the smaller plate internal to the cutout is known as the ejector plate. As the name implies, the function of the ejector plate is to help eject the finished parts from the mold. To eject the parts, the ejector plate is moved forward, into the space created by the cutout in the ejector box. Ejection of the parts is achieved through the use of a series of pins, the inserts that form the optical surfaces themselves, or a combination of the two. The pins or inserts are connected to the ejector plate, running up through the cooling plate to the core plate. When the ejector plate is pushed forward, the pins or inserts move forward as well, pushing the parts off of the core plate. When the optic-forming inserts are used to eject the parts, this is referred to as optical eject. While in conventional part molding the parts may literally be ejected from the mold, dropping into a bin or onto a conveyor beneath the mold, molded optical parts are not usually flung from the mold. Instead, a robotic arm with a clamp or vacuum suction device, known as a picker, grabs the parts. In this case, the ejection mechanism is moved such that the parts are pushed far enough off the core plate to allow any part depth to be cleared as the picker removes the parts from the mold area. In most cases the parts, gates, runners, and sprue all come out of the mold as a single piece. The runner system can be designed in a manner allowing it to be used as a handling feature for future processes, such as assembly.

The cooling plate on the moving half of the mold serves the same function as on the fixed half, though the two halves may be operated at different temperatures. The core plate has inserts to form optical surfaces on its half of the molded optic component. The core plate may or may not have runners cut into its face. If it does, the runner will essentially form a full cylinder, a condition known as a full-round runner. If not, the result is a half-round runner. In addition to the runners, shallow channels may also be seen on the faces of the core and cavity plates. These channels, called vents, allow the air in the mold to escape when the plastic is injected. If air is trapped in the mold during injection, it may burn, due to the high pressure created, or may form a small bubble that appears as an indentation on the part; in either case, the part may be ruined. Proper venting is essential to the production of molded plastic optics. This requires both appropriate vent design and keeping the vents clean and open with proper tool maintenance.

The plates comprising the molds are usually made of hardened tool steel, which allows the mold to withstand the large forces at work during the molding process. In some cases, softer materials may be used, especially for producing a limited use prototype mold. Production molds are almost always hardened steel and can be expected to have a lifetime of at least 1 million molding cycles, referred to as a class A tool. The inserted pieces forming the optical surfaces, often called optic inserts or optic pins, can be formed in several ways. The most common method is to fabricate a steel pin with the approximate surface form, nickel plate the end of the pin, then machine the nickel layer to the final optical form using diamond turning. Diamond turning is a precision machining process that uses diamond tools and a high-precision lathe to create optical quality surfaces. Usually, the diamond turned nickel surface does not even require any postpolishing. Improvements over the years have led to diamond turning machines that are capable of producing aspheric, diffractive, nonsymmetric, and free-form surfaces. Most surface forms that can be designed using optical design software can be generated using diamond turning.

Detail around the optical surfaces, including any lens flanges, is usually formed by a machined piece that is inserted into the core and cavity plates. This piece is often referred to as the cavity. In many molds the optic inserts

FIGURE 4.8
Injection molding cavity set.

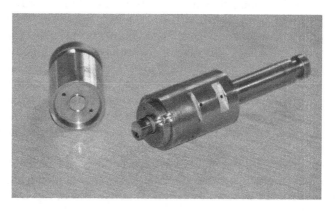

FIGURE 4.9
Injection molding cavity set, with optic insert forward for part ejection.

first go into a hole in the cavities, and then the cavities go into their own holes in the core and cavity plates. The combination of optic insert and cavity is referred to as a cavity set. A cavity set for creating a biconvex lens is shown in Figure 4.8. The cavity set half on the left of the picture goes into the fixed mold half, and the one on the right in the moving mold half. The cylinder coming from the back of the cavity on the right is an ejector rod, which gets attached to the ejector plate. The optic insert is attached to the front of the ejector rod. When the ejector plate is driven forward, the ejector rod and optic insert also move forward, pushing the lens out of the cavity detail, an example of optic eject. Figure 4.9 shows the cavity set with the insert forward, as if the ejector system was engaged. The optic pin for the other lens surface, which does not move, is visible in the fixed mold half piece. The axial position of the optic pins, which sets the center thickness and flange to vertex dimensions of the lens, is adjusted using thick metal spacers. The spacers, located in the cavity behind the optic pins, can be ground to the length needed to properly position the optic surfaces.

Alignment of the two halves of the molds in obtained through a series of features. Rough alignment is provided by "leader pins," which are the cylinders located on the outer corners on the cavity plate, as can be seen in Figure 4.6. The leader pins engage the corresponding holes in the core plate. The next level of alignment is provided by mold half-taper interlocks, which are the eight conical features seen on the cavity plate, two outward of each array lens cavity set. These protruding tapers engage the corresponding features on the core plate and, for this mold, set the fixed half to moving half final alignment. For a higher degree of alignment, the two halves of the cavity sets themselves can be taper interlocked, as seen with the cavity set shown in Figure 4.8.

Optical plastics shrink during the molding process, typically up to about 1%. Because of this, the dimensions of the details machined into the cavities are not the same as the dimensions on the part print. As an example, consider making a mold for a lens with a print diameter callout of 10 mm. If the material being molded shrinks by 1%, the diameter of the hole that is machined into the mold would need to be 10.1 mm to produce the proper part diameter. Shrinkage also applies to the optical surfaces, requiring modification of the surface form diamond turned onto the optic insert. To make matters worse, the shrinkage can be nonuniform across the part and may vary with molding process settings, such as mold temperature. In order to deal with this, molds are sometimes initially machined in a "steel safe" condition. A mold in a steel safe condition has extra material in place that can be machined away as needed. As an example, consider the 10 mm diameter lens again. If we wanted to leave the mold in a steel safe condition, we might machine a 10.05 mm diameter hole instead of going straight to a 10.1 mm diameter hole. Thus, we have a steel safe condition of 0.05 mm. After molding and measuring an initial set of parts, we would calculate how much, if any, the diameter needs to be opened up. Depending on the complexity of the part, we may want to make several iterations of molding, measuring, and adjusting. The process of adjusting the mold in this manner is known as mold compensation. The larger, more complex, and more tightly specified a part, the more likely that mold compensation will be required. This can be particularly true for relatively large, thick lenses with highly aspheric surfaces, as nonuniform shrinkage can make predicting the required optic pin surface form difficult.

4.3.3 Injection Molding Process

The injection molding process begins with preparation of the injection molding machine, the mold, and the plastic material to be molded. The plastic material is placed into a dryer, which removes water that the material has absorbed during storage. The dryer is set to a specific dew point, ensuring that the material is sufficiently dry, as excess water in the material will result in molded optics that are cloudy. The mold, having received any necessary maintenance and cleaning, is placed in the molding machine and attached to the platens. Hoses are connected between the thermal controllers and the cooling plates on the mold. Heated oil is circulated through the hoses and cooling plate channels in order to bring the mold to the appropriate temperature. Electrical connections, if any, are made; an example of an electrical connection would be for an internal mold heater. The injection barrel and screw are prepared for the molding run. If a different material was used prior to this molding run, the barrel will be purged of the old material and cleaned. If necessary, the injection screw will be swapped out. The heater bands around the injection barrel are programmed for the desired temperature profile and the heating process begun. When the material has been sufficiently dried, it is placed into the hopper on the molding machine, where it is fed into the injection barrel. The injection screw is rotated, which transports the material forward toward the mold. Between the friction produced on the material by the screw rotation and the heat from the barrel heater bands, the material is molten and at the desired temperature by the time it reaches the end of the injection screw. With the mold, injection screw and barrel, and material all at their proper temperatures, an injection cycle can begin.

An injection, or mold, cycle consists of all the activities necessary to produce a molded plastic optic and to be prepared to produce another one. The mold cycle begins by closing the mold, with the moving platen driven forward until the two molds halves are connected, with the leader pins and taper interlocks engaged. Force is applied from the clamping mechanism to ensure that the mold will remain closed during the injection process, which occurs next. To inject material into the mold, the injection screw is pulled slightly backwards, allowing material to collect at its front. This material, known as the shot, is injected into the mold by driving the injection screw forward. The molten plastic enters the mold through the sprue, passes through the runners and gates, and fills the cavities, forming the parts. Pressure is maintained by the injection screw, forcing the material to fill the cavities and preventing any backflow. This process of applying pressure on the injected material is known as packing the mold. The molten plastic, being at a higher temperature than the mold, transfers its heat to the mold plates and to the oil in the cooling channels. As the temperature of the plastic goes below its glass transition value, the material hardens. This typically occurs first at the input to the cavity, the gate. When the material in the gate hardens, it is said that the gate has "frozen off." No matter how much packing pressure is applied, once the gate is frozen, no more material will enter the cavity.

The next step in the molding cycle is to wait. Because the plastic carried a significant amount of heat with it into the mold, there will be some finite amount of time before the plastic has solidified enough to be handled. Opening the mold and attempting to remove the parts too early can produce distortion or sagging of the molded optic, resulting in the part not meeting its specifications. This waiting period, known as the hold time, is normally the longest part of the molding cycle. The hold time for plastic molded optics is long compared to similar nonoptical plastic molded components. This is because the relatively tight specifications on molded optics typically require the part to be fully hardened before it is removed from the mold.

Once enough time has passed so the part is sufficiently cooled and hardened, the mold can be opened. The moving platen is pulled back, opening the mold at the parting line and creating a space between the two halves. As the moving half of the mold pulls away, the parts that have been molded move with it, releasing from the fixed half of the mold. Having the parts pull, that is, release from the fixed half and go with the moving half, is part of the mold design. In some cases features known as pullers or grippers are put into the moving half of the mold to ensure the parts pull. With the mold open, the picker is moved into place. If the picker is using a clamp mechanism, the clamp will normally grab the sprue, which will be sticking out from the core plate. If a vacuum device is used, it will be placed against the parts and the vacuum activated. With the parts secured by the picker, the ejection mechanism is now engaged, pushing the ejector plate, the ejector pins, and the inserts forward, removing the parts from the moving half of the mold. The ejector motion is provided by the clamping motion and is controlled in order to push off the parts without distorting them. In some cases, the ejection stroke is actually a two-piece motion. First, a small "bump" is applied to the ejector plate in order to break the parts free from the grippers and the core plate detail; then a second motion is applied to push the parts from the core plate. With the ejection performed the ejector plate is returned to its original position, moving the ejector pins and inserts to their starting position as well. The parts are now no longer touching the mold, but are held in the space between the mold halves by the picker. The robotic picker arm next moves to remove the parts from the molding machine, placing them on a tray, a conveyor belt, or holding them out for the machine operator to take and inspect. With the parts out of the molding machine and the picker out from between the mold halves, we have now completed a mold cycle. To produce the next set of parts, the clamping mechanism is driven forward, the lead pins and taper interlocks engage, the mold closes, and the steps above are repeated.

The time required to perform all the steps in a molding cycle is referred to as the cycle time. The cycle time is the primary factor in setting the price of the molded plastic component. Knowing the cycle time, the number of cavities in the mold (actually the number of conforming parts created in each cycle), and the shop floor rate of the geographic region the parts are produced in allows a direct estimate of the cost of producing the base molded plastic optic component. Cycle times can range from seconds to minutes, depending on the part size, material, and specifications. In general, the tighter the requirements on the optical surfaces are, the longer the cycle time is. For a standard-looking lens with a diameter less than a centimeter, cycle times of approximately forty-five to ninety seconds might be expected. Secondary operations, such as finish machining or applying optical coatings, can also impact the piece part price.

Material costs do not generally have a large impact on part price, unless an exotic material is used or the parts are quite large. A significant number of parts can normally be made from a kilo of material, and with material prices generally in the $10 to $50 per kilo range, material costs are often dwarfed by cycle time. The amount of material used in the runner system can be significantly higher than that in the parts themselves. In some cases the runner is effectively removed by placing it internal to the mold plate and keeping the material in it molten, referred to as a hot runner system. While saving material due to not molding a runner with each mold cycle, hot runner molds can be more expensive to build and more complicated to operate than standard cold runner tools.

Using the proper settings for the parameters of the molding process is critical to optimizing the production of molded plastic optics. The parameters, of which there are many, include the injection barrel heat profile, the injection speed and pressure profile, the packing pressure applied, and the mold half temperatures. Modern injection molding machines are computer controlled, allowing precise control and monitoring of the molding parameters. The parameter values are usually determined by the mold process engineer, and the development of the values is known as processing the mold. A standard method of processing the mold is to first set the molding parameters to reasonable values for the part size, part structure, and material, and then perform a design of experiments. The reasonable values are usually arrived at from the mold processor's experience, along with recommended settings from the material manufacturer. In general, temperatures for producing molded plastic optics are higher than those for producing nonoptical components.

To begin the experiment, the processor typically will perform a series of "short shots." Short shots are mold cycles where the amount of plastic injected into the mold is not enough to fill out the mold cavities. By slowly increasing the shot size and looking at how the parts fill out, the processor can determine the flow of the molten plastic into the cavities. Ideally, the material will flow smoothly through the cavity, having a single flow front that fills evenly from the gate to the volume opposite it. If the flow splits into several pieces after the gate, rejoining later in the part, the result is normally a flow line. A flow line is a defect in the part caused by the discontinuity of the two flow fronts where they meet. The reason that most molded plastic optics use a single gate for each cavity is to limit the input to the mold cavity to a single flow front. This does not prevent a flow line from developing if the flow front splits, but ensures there is only one initial flow front to deal with. If necessary, the size and shape of the gates in the mold may be adjusted to help control the initial flow into the cavity.

Once suitable flow has been achieved and a full shot can be performed, initial shots are made and evaluated. The form of the optical surfaces is measured and compared to the requirements. Adjustments are made to the molding parameters, with more parts made and measured until a reasonable process is achieved. Removing the mold from the machine, adjustments in the location of the optical surfaces (flange to vertex depth and center thickness) are performed by machining the spacers beneath the inserts. The mold is then returned to the molding machine for final processing. In an iterative fashion, parameters are adjusted, parts made and measured, and the results fed back into the process. With a stable process achieved, the mold may be removed from the machine in order to perform mold compensation, if necessary. Once final mold adjustments and compensations are performed, the final mold process is determined and programmed into the machine.

The goal of the mold processor is to achieve a stable, efficient molding process. Since cycle time is often the largest cost driver, the mold process engineer seeks to minimize the cycle time, to the degree that conforming parts can still be delivered. It should be understood that most molding vendors consider their final process parameters to be proprietary information. If the customer wants or needs to know this information, that should be spelled out in the molding contract. Different vendors will likely develop different mold process parameters, based on their processor's experience and knowledge of the equipment and materials used.

4.3.4 Secondary Operations

Depending on the parts produced, the molded plastic optics may be subjected to a series of post-molding processes, known as secondary operations. The most common secondary operation is removal of the part from the runner system. This process is known as degating, as the part is typically cut from the runner at the gate. Degating can be performed in a number of ways, from the machine operator cutting the gate with a pair of handheld side cutters, to hot knives, and even robotic ultrasonic or laser degaters. Once degated, the parts are placed in trays, tubes, or bags, depending on the next operation, if any. For automated assembly and handling operations, the runner system may be utilized as a handling fixture, and the part not degated until the final assembly station.

Another common secondary operation is coating. Similar to glass lenses, antireflection (AR) coatings may be applied to the optical surfaces to improve transmission and prevent ghost images from surface reflections. In addition to AR coatings, there are a wide range of coatings that can be applied to molded plastic optics. These include reflective coatings for mirror surfaces, hard or antiscratch coatings that prevent surface damage, transparent electromagnetic interference coatings, water absorption prevention coatings, and laser and dichroic bandpass coatings.

While similar in performance to the coatings for glass optics, coatings for plastic optics are distinct in two ways. First, the standard method of applying coatings to glass optics cannot be used in the application of coatings to molded plastic optics. This is because standard glass coatings rely on high temperatures to activate the glass surface and produce a coating with suitable adhesion. In the case of plastic optics, the temperatures used for glass optic coating would melt or deform the molded plastic element. Instead of using heat, the surfaces for molded plastic optics may be activated in other ways, such as the use of an electron beam gun. Second, because plastic optics have a larger expansion with temperature than glass (higher coefficient of thermal expansion (CTE)), plastic optic coatings must be able to withstand significant expansion and stress. Crazing, the development of a system of small cracks, usually resulting from thermal exposure, is a common failure mechanism for poor plastic optic coatings. It is recommended that coated plastic optics, at least on a sampling basis, be exposed to thermal cycling to verify coating performance.

As a result of these differences, the coating of plastic optics is a specialty within the coating industry. Most molders of plastic optics have in-house coating capabilities or have access to coaters with significant plastic optic coating experience. It is not uncommon for plastic molders and coaters to be closely geographically located, as this reduces the shipping and handling required, and allows tight interaction between the molder and coater.

In addition to optical coatings, it is also possible to apply blackening or colored coatings to molded plastic optics. This may be to eliminate stray light paths or to enhance the cosmetic appearance of the part. In some cases, these black coatings can be applied by painting or ink stamping the parts. In other cases, the parts may be dipped or sprayed. Adhesive backed films may also be used.

Depending on the complexity and production volume of the part, machining operations may also be performed as a secondary operation. Secondary machining may occur in situations where it is easier or cheaper to mold a simple part and machine it than it is to mold a complex part having all the desired features. Secondary machining sometimes occurs for a short period during initial production, when a design change is required and before the mold can be reconfigured, or a new mold created. Because secondary machining may require significant handling of the part, there may be an associated yield loss with it.

4.3.5 Tolerances

The tolerances that can be achieved in molded plastic optics are mainly driven by the injection molding process, the quality of the mold that the parts are produced from, and the amount of time and effort placed in achieving them. Table 4.2 shows tolerances that are readily achieved in the production of molded plastic optics. Some physical dimensions, such as center thickness and vertex to flange distance, can be adjusted in the mold using precision ground spacers. Other dimensions, such as lens diameter, are typically controlled by the machining of the mold. In the case of lens diameter, the hole within the mold and the shrinkage of the material in the molding process

TABLE 4.2

Standard Tolerances for Injection Molded Plastic
Optics (Tighter Tolerances Achievable)

Radius (%)	±0.5
Focal length (%)	±1.0
Diameter (microns)	±20
Thickness (microns)	±20
Surface form (fringes per cm)	1–4
Surface irregularity (fringes per cm)	0.5–2
Surface roughness (RMS angstroms)	20–40
Surface radial decenter (microns)	20
Surface quality (scratch/dig)	40/20
DOE microstructure depth (microns)	±0.1

Note: Fringes are at 632.8 nm.

will set the final dimension. Alignment between the two halves of the mold, and therefore the features contained in them, is controlled through taper interlocks, as discussed above. Surface radius and form error can be adjusted both through the mold process parameters and through tool compensation. Surface form errors can also be reduced by longer hold times, which equates to a higher cost.

The tolerances in Table 4.2 are collected from data provided by various molders. Tighter tolerances can and are regularly obtained. For instance, in the design of cell phone camera lenses, a surface centration of 20 microns is unacceptable. For these elements, decentrations of less than 5 microns are required and are being consistently achieved. While we recommend that designers always talk with molders early in a project, this is particularly true when the design requires tolerances that are tighter than shown in the table. Working with a molder early in the design process can guide decisions, including tolerance limits, which may affect the particular design form chosen.

Performing a tolerance analysis on a molded plastic optic design is similar to the tolerance analysis for a glass optic system. The basics of tolerance analysis were discussed in Chapter 1 and are not repeated here. We do note that selection of the probability distributions for the various tolerances can make a significant difference in the yields and performance predictions that the analysis produces. Because molded optics are produced by replication, they are not necessarily uniformly distributed with regard to their parameter values. We recommend using several tolerance distributions in the analysis, including endpoint and parabolic, to consider the potential worst-case scenarios.

4.3.6 Injection-Compression Molding

We stated earlier that standard injection molding machines do not press directly on the material in the mold; in injection-compression molding, the machine does just that. Unlike a standard injection mold, where the optic insert is designed to be in a fixed position, an injection-compression mold is designed to allow the optic insert to be moved within the tool during the mold cycle. The ability to press on the injected material in a controlled fashion provides another degree of freedom to the mold processor. Injection-compression molding is often used to produce parts that are not easily made using standard injection molding. This includes wide, thin parts, as well as parts with large thickness variations or optics requiring tightly controlled flat surfaces, such as prisms. For wide, thin parts it may be difficult to inject the plastic and have it flow through the entire part, resulting in voids in the outer part regions. With injection-compression molding, the mold inserts can initially be held apart at a thickness greater than that of the final part, then pressed to the final dimension after the plastic has flowed throughout the cavity. If needed, overflow channels can be utilized to allow more plastic in the cavity region than is necessary to fill out the part. For parts with large thickness variation, such as a prism, nonuniform shrinkage may make it difficult to produce consistent parts using standard injection molding. In this case injection-compression may help to control the shrinkage by moving the insert during the hold time of the molding cycle.

Injection-compression molds and molding machines look very similar to standard injection molding machines and molds. However, the added complexity required to move the insert during the molding cycle (and not just to eject the parts) generally results in a more costly mold. Most injection-compression molders also perform standard injection molding, while the opposite is not usually true. Injection-compression, because of its added complexity and cost, is normally used when standard injection molding will not suffice.

4.4 Design of Molded Plastic Optics

In many ways the design of molded plastic optics is similar to the design of optics using other materials, such as glass. The use of molded plastic optics, however, does bring some unique design considerations with it.[2–6] Designers familiar with developing systems with glass optics should find the transition to molded plastic optics relatively straightforward, though they may want to unlearn a couple of glass optical design rules of thumb.

4.4.1 Design Considerations

As with most optical designs, the design of a molded plastic optic system ideally begins with a well-defined set of requirements. In reality, such a requirement set is often not available, though general desires and guidelines are understood. For instance, if developing a design for a cell phone camera, it is unlikely that there is no knowledge of the desired field of view. Instead, the customer will have some range of field of view it considers acceptable. In this situation, the designer may perform a trade study to determine the effect of field of view on the design. Along with a general understanding of the base properties of the system, such as field of view, F/#, and image quality required/desired, the most important requirements when developing a plastic optic design are environments and cost.

Environments can make a critical difference in the plastic optic materials that are selected for the design, or may even determine that molded plastic optics are not a suitable design choice. The primary stressing environments for plastic optics are thermal conditions. Thermal conditions affect molded plastic optics in two ways. First, if they are too severe, the plastic optic may melt or soften, deforming the optical surfaces and degrading the system performance. Understanding the severity of the thermal environments allows selection of the appropriate plastic optic material, for as we saw earlier, the various materials have different service temperatures. When considering thermal environments, it should be kept in mind that the transportation/storage environment may be more thermally stressing than the operating environment. Second, assuming the material can survive the thermal environment, we must deal with the fact that plastic optic materials have large changes in index with temperature (dn/dt) and relatively large thermal expansion (CTE). If not properly accounted for or compensated, these changes can result in a large focus shift over temperature.

Chemical environments can also have a severe effect on molded plastic optic systems. It is well known that acetone, a chemical commonly used in cleaning glass optics, can degrade plastic optic materials. Unusual environments, such as military settings, where an optic may be exposed to jet fuel fumes, deicing solutions, etc., must be factored into material selection. While many commercial product applications do not see these types of exposures, understanding the expected chemical environment is imperative, as it may cause catastrophic failure of molded plastic optic system performance.

Cost is another critical requirement to understand during the design of a molded plastic optic system. Cost is often the primary reason for the consideration of utilizing molded plastic optics. Since the cost of the optical portion of the system is directly related to the number of optical elements in the system, cost requirements may limit the number of lenses allowed in a design. Cost vs. performance trades are a common occurrence in the development of molded plastic optic designs. Again, using the example of a cell phone camera, the customer may want to see the predicted performance for two- and three-element camera lenses. The customer can then determine whether the 50% increase in the total cost of the lens elements is worth the performance improvement created by having the third lens in the system.

With the requirements understood, or at least bounded, the design of the optical system can begin. The design process typically begins with the determination of what materials are available for use in the system. If the temperature environments are too high, it may be that acrylic cannot be used, but cyclic olefin polymers are acceptable. Similarly for the flint materials, it may be that polystyrene is not suitable, but polycarbonate is. Since there are a fairly limited number of materials to choose from, determination of the available materials is usually a straightforward task.

With material choices determined, design form is often considered next. In the design of glass optical systems, there is a wide range of "classical" design forms to choose from. Systems such as the Cooke triplet are well known and understood. There are several texts that provide examples and descriptions of classical glass optic designs.[7,8] For plastic optic systems, there is less of a public database of design forms. This is in part because of the shorter history of use of plastic optics, but also due to the proprietary nature of the business, as well as the limited number of highly experienced plastic optic designers. Most companies that design or produce systems using molded plastic optics have an in-house database of prior designs. These are often used as starting points for new, variant designs. For those without access to such a database (which are most designers of plastic optics), the best places to find design starting points are the patent literature, conference proceedings, and comparable products currently on the market, if any. If a designer works a significant number of projects involving plastic optics, they will find that they naturally create a database of designs, as well as developing intuitive knowledge of what a good starting point may be for a given application. If no starting points are readily available, the simplest method is to put in some reasonable-looking lenses and begin optimizing them. For example, if a two-element lens is desired, place a positive crown lens and negative flint lens into the design software, use a large valued F/# (stop the system down) and a reduced field of view, and start from there. When the design is under control and optimizing, the F/# can be lowered and the field increased until the desired values are obtained.

Thought should be given early in the design process to the optomechanical features of the system, along with their interaction with the optical design. Integration of flanges, mounting tabs or snaps, alignment pins, bosses, and antirotation or orientation features is one of the advantages of using molded plastic optics. Flanges can be used to generate the correct spacing between lenses or to protect optical surfaces from touching a surface if the optic is set down, while alignment pins can ensure proper orientation of elements. Features can also be created for automated or manual handling or assembly, such as dot patterns for vision systems (human and robotic). Interaction between the optical and optomechanical designer should occur throughout the development of the design, to help fill the needs of both.

Many molded plastic optic design tasks are heavily constrained; that is, there are many conditions imposed on the system. In a cell phone lens, it may be that for cost reasons, only three lenses are allowed, the overall height from the front of the lens barrel to the image plane must be less than 6 mm, the angles of incidence on the image plane must be less than 25°, and the image performance must be relatively uniform over the field. Under these conditions, there will be a limited number of design forms that can be utilized. The goal of the designer will be to make the design form selected as cost-effective and producible as possible.

Creating a cost-effective, producible molded plastic optical design requires a basic understanding of the component manufacturing method, which was described above. As noted earlier, cycle time is the primary factor in setting the cost of plastic optics. The designer can help to minimize the cycle time, and thus the cost, in two main ways: part thickness and part specifications. Thicker parts take longer to cool than thinner parts, as the larger quantity of plastic used to produce them carries a larger heat load into the mold. Having more heat to remove from the part increases the hold time, which makes up most of the mold cycle time. Of course, there is a limit to how thin the parts can be made, which is discussed below. The basic rule is that the part should be only as thick as needed to both perform its function and be moldable.

Obtaining suitable element thickness is best achieved by constraining the maximum thickness allowed in the lens design code. Left to their own devices, lens design programs will often "fill the world" with plastic; that is, they will continually increase the thickness of the elements until all the design space is plastic. This occurs because the design code determines that having thicker elements provides an improvement to its merit function. From the designer's and molder's perspective, the improvement may or may not be worth the increase in cycle time associated with the thicker part. Unless told to, however, the design program does not consider this. By constraining the allowed thickness to reasonable values, the designer should be able to develop an initial design that can straightforwardly be produced. To determine the benefit from increased thickness, the maximum thickness constraint value can next be increased or removed. Reoptimizing the design, the performance improvement relative to the thickness change can be evaluated.

In addition to overall part thickness, the amount of thickness variation that a part has can also affect the cycle time. In theory, a part with uniform thickness is ideal to mold, as the shrinkage should be quite consistent across the part. However, uniform thickness parts generally do not provide the optical power needed for most designs. An example of a powered element with a significant thickness variation is shown on the left in Figure 4.10. This biconcave lens has a center thickness much larger than its edge thickness. The shrinkage across the part is likely to be highly nonuniform, potentially requiring mold compensation. Depending on the application and requirements, it may be more cost-effective to split the element into two pieces, as shown in the right of the figure.[9]

Part specification can also drive the length of the cycle time, by affecting the required hold time. Parts with tight radius, surface form, or surface irregularity requirements will generally need a longer hold time than

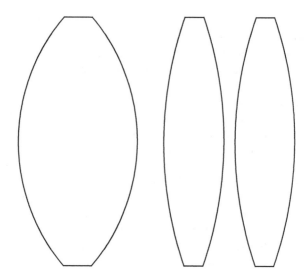

FIGURE 4.10
Biconvex lens with significant thickness variation (left), converted to more moldable pair of lenses (right).

components with looser specifications. Thus, setting an unnecessarily tight surface specification can result in increased part cost. A common oversight during the tolerance analysis of molded plastic optics is to consider the effect of radii variation of the individual surfaces, but not of the element as a whole. It may be that the system performance appears sensitive to radius variation when each surface is independently perturbed, but the performance is not significantly degraded when both surfaces are perturbed together. That is, if both radii grow longer, or both radii are shorter, they may compensate for each other and produce little performance drop. Particularly, this may be seen in systems that have a small field of view, where the beams incident on the surfaces having little "beam wander" (motion of the beam on the optical surface as a function of field angle). Compensation of the two surfaces of a lens is an important consideration in setting the surface requirements. It is not unusual for parts to want to "spring" if removed from the mold with a shorter hold time. This type of lens springing tends to influence both sides of the element equally, so that the surface radii values move together. If the tolerance analysis shows that surface compensation would allow adequate system performance with a sprung lens, the designer should consider specifications that allow the part to be produced this way.

Specifying the allowed springing in a lens can be somewhat complicated and requires clear communication between the designer and molder. One method of specifying the element is to allow the radii to vary, over a larger range than would be considered from a single-surface tolerance analysis, as long as the wavefront transmitted by the element meets a certain requirement. Another method is to provide a large-radius variation tolerance, but require the departure of the two surface radii from their design values to be closely matched. For example, if one surface has a radius that is 2% longer than its nominal value, the other surface must have a radius value between 1.5 and 2.5% longer than its design value. In this way, the two surfaces are forced to move together. An alternate method would be to specify an allowed change in power difference between the two surfaces. Whichever specification method is chosen, it should be discussed in detail with the molder, so that there is no disagreement as to what constitutes an acceptable part. In addition to changes in radius, it may also be acceptable to allow a cylindrical change in the element. This can be considered equivalent to accepting a higher level of irregularity on the part. As with the compensating change in radii, a compensating change in cylindrical power must be evaluated and specified carefully.

While reducing the thickness of a part can help to decrease its cycle time and cost, there are certain thicknesses that cannot be too small if the part is to be molded properly. This is especially true of the edge thickness of the part. Because the injection molding process requires the molten plastic to flow into the mold cavity, there must sufficient room for an appropriately sized gate. The gate size requirement translates into an edge thickness requirement on the element, at least in the area neighboring the gate. Edge thickness can be controlled during the design process by setting a constraint in the optical design code. In some codes, there is a defined edge thickness constraint. In other codes, it may be necessary to construct such a constraint. The constraint can be formed as the difference between the center thickness of the part and the sags of the surfaces, with care given to the sign of the sag. However the edge thickness is constrained, it is important that the definition match the need of the molder. If the edge thickness is calculated from the ray heights, it is assumed that the optic surface will be good out to at least this height. In reality, the molder needs room outside the used portion of the surface to account for "edge break," as discussed below. Thus, the true part edge thickness must not be based only on ray positions, but also on the ray positions plus a buffer zone. This buffer zone can be included in the definition of the edge thickness constraint, which results in this buffered edge thickness being the appropriate parameter to control.

Any part dimension that restricts the flow of the injected plastic is a potential impediment for successful production of the part. As we discussed earlier, the goal of the mold processor is to have a single flow front that fills evenly from the gate to the volume opposite it. If the center thickness of a meniscus or biconcave part is too narrow, the plastic will flow around the central region instead of through it. When the rest of the part is full, the plastic will finally be forced through the center. This is illustrated on the left of Figure 4.11. This results in a flow line, due to the combination of multiple flow fronts. To avoid this, the center thickness should be increased to allow proper flow, as shown on the right of the figure. Flow restriction can also occur in thin-walled sections or flanges. Thus, long, narrow flanges should

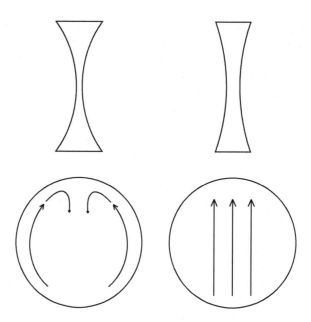

FIGURE 4.11
Biconcave lens with thin center restricting flow (left) and more suitable biconcave form for molding (right).

be avoided. If the flanges designed to provide spacing between two lenses are too long, it may be necessary to shorten the flanges and use a separate, possibly molded, spacer.

In addition to leaving adequate edge thickness for the plastic material to flow through, it is also necessary to leave an adequate radial distance between the edge of the optical surface and the clear aperture of the optical surface. The clear aperture of the optical surface is the area over which the surface requirements must be met. The clear aperture cannot generally be as large as the optical surface due to a phenomenon known as edge break or edge roll. Edge break is the distortion of the surface in the region near a transition. This transition may be between the optical surface and the edge of the part (for a flangeless element), or between the optical surface and the flange. The amount of room needed to account for edge break will depend on the part configuration and size, but keeping a 1 mm wide annular ring around the optical surface is generally a good practice.

In addition to these constraints, the design of molded plastic and glass systems differs in a couple rules of thumb. In particular, the use of planar surfaces is encouraged in glass optical designs, but discouraged in molded plastic optics. As an example, when a lens design results in a surface with a long radius of curvature, such that the optic surface has only small sag, the glass optic design rule of thumb is to turn it into a planar surface. For conventionally fabricated glass optics this makes sense, as a planar surface is easier to produce and test. For molded plastic optics, the opposite is true. Planar surfaces tend to warp during the molding process, due to a lack of self-support. The result is typically a surface with some long radius, or with irregularity. The rule of thumb for plastic molded optics is thus to shorten the radius, which provides the surface with self-support, and makes it easier to produce.

Another rule of thumb in the design of glass optics is to make surfaces that have similar radii values the same radius value, or to use surface radii values that match a specific "test plate." Many optical shops have a catalog of test plates, spherical reference surfaces that are used to test glass optics during their production. In the production of molded plastic optics, there is no test plating, and since the surface will be replicated from a custom-produced insert, there is no need for a specific radius value.

This also applies to the use of symmetric elements, which is sometimes encouraged during glass optic design. The reason for this is that a symmetric lens cannot be inserted backwards into the system. For plastic optics, the lens usually has some form of flanging or other feature that can be used to orient the lens. Thus, having a symmetric lens (same radii on both surfaces) on a molded optic with asymmetric flanges is not usually a benefit.

In addition to these general rules, the discussions regarding aspheric and diffractive surfaces in Chapter 1 apply. The use of molded plastic optics allows the utilization of these surfaces, which in turn requires appropriate considerations during the optical design. Designing surfaces that cannot be created, whether out of plastic or another material, is a poor design choice.

4.4.2 Cell Phone Camera Example

We have mentioned cell phone camera lenses throughout the chapter and now briefly examine one such design. The example used is the same system seen in Chapter 3, which was analyzed for its stray light performance. The optical design is shown in Figure 4.12. The system is approximately 6 mm from barrel front to image plane and consists of three molded plastic optics, each made of Zeonex 480R. The lens material, a COP, was selected for its service temperature, based on the thermal environments of the application. The first element is a positive lens, the second element a negative lens, and the third element a low-power aspheric corrector used for aberration control. The planar object behind the third element is the detector cover window (glass). The use of both a positive and a negative element helps with control of the Petzval curvature, as mentioned in Chapter 1. The aperture stop is at the front of the system, which eliminates the possibility of controlling coma, distortion, and lateral color through symmetry. Instead, these are controlled through the use of aspheric surfaces and a kinoform. The location of the aperture stop is based on the constraints of overall system length and maximum ray angle of incidence on the image plane. All of the elements

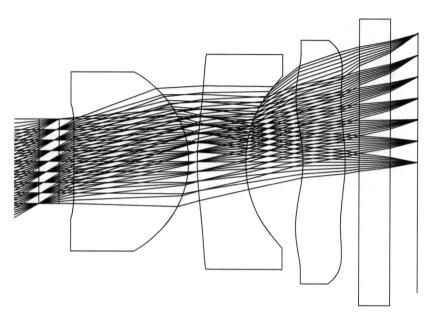

FIGURE 4.12
Optical design of three-element cell phone camera lens.

are aspheric, and the second element has a diffractive on the front surface of the lens, with a minimum grating period of 19 microns. This diffractive, positioned away from the aperture stop, is used to correct axial and lateral chromatic aberration. The diffractive is placed on a surface that has been designed to be relatively shallow, aiding in its production. The surface was also selected because it has a relatively low variation of angles of incidence over it (compared to the other surfaces), which helps to optimize the diffraction efficiency.

The shapes of the elements lend themselves well to molding, as there are no significant thickness variations, as well as no significant flow restrictions. The optical surfaces have been extended outside the area used by the rays, in order to account for edge break in the molding process. The gull-wing shape of the final element is seen quite often in designs for this application. This form is being used to control the angles of incidence of the beams onto the detector, particularly near the corners of the field. It also can be used to control distortion, which for this design changes sign across the field of view, varying from +2 to –3%. Since the aperture stop is located at the front of the system, it can be molded into the barrel that will house the elements. The use of flanges for lens spacing works quite well in this design. One possible flange configuration is shown schematically in Figure 4.13, where the flanges are designed such that each lens has at least one flange that extends beyond the maximum sag of the surface. This allows the element to be set down on a tray, while still protecting the optical surface. A spacer has been

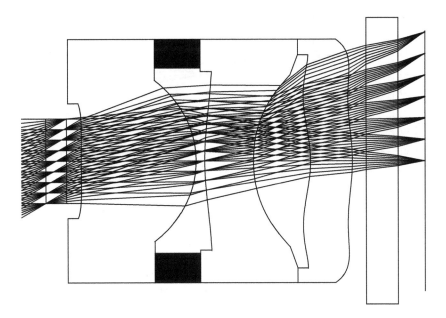

FIGURE 4.13
One possible flange configuration for the cell phone camera lens.

placed between the first and second elements in order to maintain short flange lengths. The spacer can be roughened to reduce stray light artifacts resulting from reflection off its inner surface. Alternatively, the spacer could be removed by having a barrel with stepped internal ledges, instead of a barrel with a constant inner diameter. The design can be easily assembled by loading the elements from the rear into the single-piece barrel.

The nominal Modulation Transfer Function (MTF) performance of the design is shown in Figure 4.14. We can see from the plot that the nominal design provides good image quality over the entire field of view. A simulated image for the nominal design is shown in Figure 4.15. Some darkening in the corners of the image is visible in the figure. This is due to the reduced illumination in the corners, which is a result of some ray clipping (vignetting) as well as "cosine to the fourth" effects. For this application, the illumination drop-off can be dealt with in the detector gain, electronics, and image processing software.

Of course, it is the as-built performance of the system that ultimately matters. We have mentioned earlier that standard injection molding decenter tolerances do not suffice for this application. As an example of this, a simulated image for the design is shown in Figure 4.16, where a 10-micron decenter has been applied to each surface. The front surface of each element was moved upwards, and the rear surface of each element shifted down. Comparing the image with surface decentrations to the nominal design image, we can see an obvious loss of image detail. This is due to the aberrations induced by

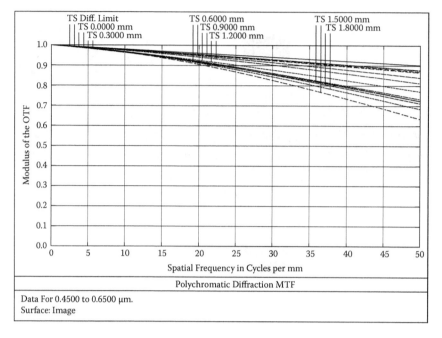

FIGURE 4.14
Plot of the MTF for the nominal cell phone camera lens design.

FIGURE 4.15 (See color insert.)
Simulated image for the nominal cell phone camera lens design.

FIGURE 4.16 (See color insert.)
Simulated image for the cell phone camera lens with 10-micron surface decentrations applied.

decentration of the surfaces. While in this example we have only applied surface decentrations, a real as-built system would also have decenter and tilt
of the optical elements themselves, as well as radius, thickness, and surface
form errors. These additional errors would further degrade the image. As
a result, surface decentrations of elements for this application are typically
specified at no more than 5 microns. Reducing the centration tolerance from
10 microns, as in the simulated image, to 5 microns provides image degradation allocation to the other tolerances.

4.4.3 Prototyping

Once a design form has been selected and developed, it is common to prototype the design before moving to full production. This is due to the significant cost and schedule associated with building, processing, and possibly
compensating a production mold. In addition to providing evaluation of a
design's performance and ease of production, prototyping also creates the
opportunity for the product to be shown to potential customers. This can
result in early feedback, while changes are still relatively easy to insert into
the design.

Planning is an important part of the prototyping of a design. There should
be plans in place that describe how the prototype parts will be handled,
assembled, tested, and disassembled (which may turn out to be necessary).
Often, there are a fairly limited number of prototype parts available, so it

is important that they be dealt with carefully. This includes the individual components as well as any assemblies they are used to create.

Prototyping of a plastic optic design is usually performed in either or both of two ways—machining and molding. In the case of machining, diamond turning is the most common method of prototyping. To diamond turn a prototype plastic optic element, cast, extruded, or molded plastic is first rough machined into blanks. These blanks may contain features such as mounting flanges and alignment tabs, or they may be simplified versions of what would be in the production design. The blanks are next placed on the diamond turning machine, where the optical surfaces of the element will be produced. Various fixturing and alignment aids are used to properly position the blanks on the diamond turning lathe's spindle. In some designs, it is possible to diamond turn the element centration feature, e.g., the diameter, at the same time the optical surface is machined. This allows the surface and centration feature positions to be tightly controlled with respect to each other. When the optical element contains two opposing surfaces, as in a standard lens, machining the diameter with one surface allows it to be used as a precision centering feature when the second surface is machined.

The precision of diamond turning machines has developed to a point that postpolishing of plastic optical surfaces is no longer required. The prototypes come off of the machine ready to use (unless they need to be coated). Most optical plastics diamond turn well, with surface roughness similar to what is produced by molding. The exception to this is polycarbonate, which tends to be a gummier material than the other optical plastics. Diamond turned polycarbonate surfaces may have a surface roughness that is higher than that produced by molding. PES and PEI can also be more difficult to diamond turn, though processes for these materials have been developed, and prototypes are regularly machined from them. Diamond turning can produce a number of different surface types. Standard, rotationally symmetric surfaces are readily manufactured, as are aspheric surfaces, diffractive surfaces, and Fresnels. The use of tool servos on the diamond turning machine allows the production of surfaces that are not rotationally symmetric, such as cylinders or free-form surfaces. Most plastic optic molders have diamond turning capability in house or can facilitate access to a diamond turning facility.

Molding is another method of producing prototypes of plastic optics. Similar to prototypes manufactured by diamond turning, molded prototypes may contain all of the features of the final design or may be simplified versions of the elements. The amount of detail in the prototype may influence the time needed to produce it, as well as which mold type is used. Molded prototypes may be created from either stock or custom molds. Stock molds are often nothing more than molds with cavity sets having holes of varying diameters. The cavity set with the diameter nearest to or oversized from the design can be used to produce a prototype element. With a stock mold, creating a molded prototype element requires production of optical inserts for the mold, but does not require production of a mold base or cavity set. Because only the insert is produced, the amount of detail that can be put into the element is limited. Some detail can be machined into the optic insert, but this can result in the part sticking to it, making it more difficult to meet the part specifications.

Custom molds allow a higher level of detail to be produced in the prototype, but require machining of the cavity set as well as optic inserts. While potentially taking longer to create, a custom mold with all of the detail of the prototype allows evaluation of the actual design. In addition, molding of the detailed prototype element will assist the molder in developing the production molding process. Evaluating the actual design can be critical, particularly when stray light is a concern.

Sometimes, the custom mold for the prototypes contains only a single cavity. In other cases, where multiple elements must be produced, as in a three-element cell phone camera, the mold may be designed as a "family mold." A family mold contains a cavity for each of the three lenses. Even though all three lens cavities are in the mold, the three lenses are generally not produced at the same time, as the mold process parameters will vary between them. Also, if the elements are different sizes or volumes, the mold will be unbalanced.

Prototype molds may be made of softer materials than tool steel, which is regularly used for production molds. Softer materials are easier to machine, but will have a limited lifetime. If it is expected that the design will be going to production, the prototype mold may instead be produced from production quality tool steel. In this case, the mold may be developed as a pull-ahead mold. A pull-ahead mold is one where holes for multiple cavities (of the same lens) are machined into the mold plates, but the cavity sets for them are not developed until needed. For prototyping, only one or two cavities may be made. Once the product goes into production, more cavity sets can be machined and inserted to the mold. In this way, the prototype parts are identical to the production parts, and evaluation of the prototype should give a good estimate of the final production performance.

In addition to prototyping optical elements, it is often necessary to also prototype mechanical parts of the design, such as lens barrels. If plastic, these prototype parts can be created in a manner similar to that of the optic elements: they can be machined or molded. If metal, the parts will likely be conventionally machined. If the prototypes of the optical elements have been made as simplified versions of the final design, the mechanical parts may need to be adjusted to compensate for this. Depending on the prototype parts' flanges, the use of additional spacers may be necessary.

A word of caution in regard to prototypes is to evaluate their performance with respect to the production quality they represent. Diamond turned optical surfaces likely have less surface form error than molded optical surfaces. Machined aluminum barrels potentially control element centration better than molded plastic barrels. Thus, a prototype made with diamond turned lenses and machined aluminum barrels may show imaging performance that is superior to that produced by most of the production systems the prototype is meant to represent. This can "set the bar too high," resulting in unrealistic customer expectations. We recommend using the tolerance analysis and image simulation features of optical design software, as discussed in Chapter 1, to evaluate the range of predicted performance expected to be produced. Knowledge of the characteristics of the prototype system can allow comparison of the software predictions to the actual prototype performance.

4.4.4 Production

Once a design has been prototyped and evaluated, it (hopefully) will be moved to production. Moving a design to production requires several decisions to be made. The first decision is where to produce the design. As discussed earlier, cost is often a factor in using molded plastic optics, and the cost of a part is related to the shop floor rate in the region the part is produced. This has led many parts, particularly those for consumer devices, to be produced in low-wage regions. Other factors, such as licensing for medical products, or security issues for defense products, may override decisions based strictly on cost.

Once the production location is determined, mold issues must be resolved. In particular, determination of the number of molds, and the number of cavities in each mold, is needed. The number of cavities and molds is usually based on the expected production volumes required. The production capability of a mold is easily calculated using the cycle time, number of cavities, and number of shifts that the mold will be run. When large numbers of cavities are needed, a decision as to the number of molds may be needed. For instance, if sixteen cavities are required to produce the number of parts needed, one sixteen-cavity or two eight-cavity molds could be built. The decision to use one or two molds will depend on a number of factors, such as machine size availability of the molder, cost of the additional mold, and risk aversion (if one mold breaks in the eight-cavity case, there is still a second mold to use for part production).

In addition to part production, thought must also be given to packaging and shipping of the parts. Appropriately designed packaging can prevent the molded optics from damage during shipping. In general, protection of the optical surfaces is desired. A variety of tubes, trays, and bags are used in the shipping of molded plastic optics. If the design company will be receiving components and building assemblies themselves, it must be prepared to handle the parts. As an example, molded plastic optics tend to easily become charged with static electricity; this can result in them attracting dust or particles, including small pieces of plastic generated during the degating process. The use of ionizing fans or guns is common in the handling and assembly of plastic parts, and the customer will want to obtain such equipment if it will be handling the molded plastic elements.

Production testing plans and equipment should also be put in place. For optical components, this can include the parameters that will be tested, the sampling frequency for testing, and the data reporting that is required. If any specialized test equipment is needed, it should be developed and approved before the production run has begun. Calibration pieces or "golden units" should also be available before the start of production. The use of commercially available test equipment, when feasible, is highly recommended.

4.5 Future of Molded Plastic Optics

The future of molded plastic optics appears bright, as new materials, applications, and products continue to emerge. The development of COC and COP materials, designed specifically for optical purposes, is a relatively recent event, and the development of new materials is ongoing. Creation of new higher-temperature materials has occurred in response to the desires of the consumer electronics industry for optical plastics that can survive solder reflow, which is now a standard manufacturing process.[10] The desire for higher refractive index materials, as well as for materials that transmit in alternate wavebands (e.g., the IR) or which can withstand punishing environments (such as those seen in solar power applications), will continue to push material development.

Illumination, with the replacement of conventional lighting by solid-state lighting elements, is expected to be an area of growth for molded plastic optics. The unusual surfaces and part geometries required by modern illumination designs will continue to drive plastic optic production technology forward. Commercial products, such as cell phone cameras, are pushing the boundaries of what tolerances can be held in molded plastic optics, as well as what size parts can be produced. Systems combining molded plastic optics with molded or conventionally produced glass optics are increasingly being seen. These systems utilize the advantages of both materials to create systems that outperform equivalent single material designs. We expect to see more of such systems, which utilize molded plastic optics to reduce cost and improve performance, as the field of molded plastic optics continues to evolve.

References

1. Rossi, M., T. Ammer, M. T. Gale, A. Maciossek, and J. Söchtig. 2000. Diffractive optical elements for passive infrared detectors. *OSA Trends in Optics and Photonics* 41.
2. Welham, B. 1979. Plastic optical components. In *Applied optics and optical engineering*. Vol. VII. New York: Academic Press.

3. U.S. Precision Lens. 1983. *The handbook of plastic optics*. Cincinnati: U.S. Precision Lens.
4. Lytle, J. D. 1995. Polymeric optics. In *Handbook of optics*. Vol. II. New York: McGraw-Hill.
5. Baumer, S. 2005. *Handbook of plastic optics*. Weinheim: Wiley-VCH.
6. Schaub, M. P. 2009. *The design of plastic optical systems*. Bellingham, WA: SPIE Press.
7. Smith, W. J. 2004. *Modern lens design*. New York: McGraw-Hill.
8. Laikin, M. 2007. *Lens design*. Boca Raton, FL: CRC Press.
9. Beich, W. S. 2010. Polymer optics: A manufacturer's perspective of the factors that contribute to successful programs. *Proceedings of SPIE* 7788:4.
10. www.suncolorcorp.com.

5

Molded Glass Optics

Alan Symmons

CONTENTS

5.1 Introduction

There are many processes for the manufacture of glass-based products that can be defined as molding. Glass slumping, where glass is simply heated and allowed to form or slump over a mold, is the process used to manufacture craft-type products such as bowls and platters. Glass reflectors for use in lighting applications are manufactured by heating and pressing a glass gob with both male and female molds. Glass containers or bottles are manufactured by blow molding a preform of glass. Yet, none of these processes are defined as precision glass molding. These molding processes are not precision processes and generate products that do not shape or transmit light in a precise manner.

The precision glass molding process is one in which the primary intent is to form the glass into an optical component or lens, in which the intended use is the shaping or transmission of light. Precision glass molding is a high-temperature, compression molding process conducted in a controlled environment using optical quality molds to manufacture optical quality components or lenses.

The driving principle behind the development of precision glass molding technology was to develop a high-volume, low-cost method to replicate aspheric surfaces on glass. Since the generation of the aspheric surface is the most expensive part of the lens-making process, molding ensures the replication of aspheric surfaces at low cost. Generating a single aspheric surface on a mold is significantly less expensive than generating aspheric surfaces for every lens manufactured. For the very same reasons, precision glass molding is well suited for high-volume manufacturing. The technologies used to generate aspheric surfaces—single-point diamond turning, precision grinding, magneto-rheological finishing, etc.—all require high-cost equipment designed for low-volume manufacturing. Molding equipment is relatively inexpensive in comparison, especially on a cost per lens basis. The cost advantage is not as prevalent for standard spherical surfaces because the traditional and mature grinding and polishing process has become very low cost, especially in the Asian markets. Only under rare circumstances and current pricing structures is a precision glass molded spherical lens, a more cost-effective alternative to a ground and polished lens. The justification for a precision glass molded lens vs. a ground and polished lens essentially comes down to the optical design and the advantage of aspheric surfaces. The advantages of aspheric surfaces are well documented and were reviewed in Chapter 1.

The principles under consideration when evaluating a product for precision glass molding are basically the same principles under consideration when one contemplates manufacturing a lens for plastic injection molding: shape, size, volume, and cost. Hence, the use of precision glass molded lenses should really be evaluated against plastic injection molded lenses. Both methods can provide aspheric surfaces as easily as they can provide spherical ones. So in many regards the main design argument for molding vs. traditional manufacturing is aspheric vs. spherical. There are other secondary design factors that can also be considered, such as molded-in features, but the optical design is the driving factor for most products.

When one compares plastic injection molded lenses with precision glass molded lenses, plastic almost always prevails on cost. There has, however, been significant effort in the reduction of cost for very high-volume precision molded glass, primarily the development of gob preforms and transfer molding. It is interesting to note that despite the existence of precision glass molding (PGM) since the 1940s and its tangible commercial presence since the 1970s, most literature touting the advantages of plastic optics completely ignores precision glass molding.

5.2 History of Precision Glass Molding

The precision molding of glass dates back to at least the 1940s. Julian Webb of the Eastman Kodak Company patented an "apparatus for molding lenses"[1] in 1946. In his patent, Mr. Webb also touts his "improved machine" and "improved methods" and refers to "known machines," indicating that there is prior history that predates even this timeframe. Mr. Webb's apparatus is a basic precision glass molding machine. His machine exhibits all of the basic systems that will be described later in the section on press molding equipment and can be considered a precursor to modern equipment.

Historically in the United States there were two primary companies that pioneered the development of glass molding technology: Corning Glass Works and Eastman Kodak Company. Eastman Kodak showed a renewed interest in glass molding equipment in the mid to late seventies,[2-5] and Corning Glass Works began patenting moldable glasses and processes for molding glass in the early 1980s.[6,7] Both companies would continue their development

FIGURE 5.1
Early Corning glass molding press. (Courtesy LightPath Technologies.)

of aspheric molded lens products for many years. Corning would eventually have an offering of at least six molded aspheric lenses as early as 1989. Corning would eventually sell the business to Geltech, Inc., now LightPath Technologies, Inc. An early Corning glass mold press is shown in Figure 5.1.

Outside of the United States glass molding was developed primarily in Japan. The main contributors were Japanese camera companies that were fulfilling their internal demand requirements for camera systems. The history of Japanese molding is difficult to discern due to this self-sufficiency. More recent history is easier to document. In 1990 Hoya Optics had a standard offering of eight pick-up lenses and five collimating lenses. Hoya also advertised the ability to manufacture lenses up to 30 mm in diameter for "facsimiles, VTR Cameras and Optical Disks."[8] Toshiba Machine released its first commercial glass molding machine in 1987,[9] and Ohara Corporation patented a gob manufacturing process in 1981.[10]

5.3 Moldable Glass

There is much discussion on what the definition of glass should be. In ASTM C-162-92 it is defined as "an inorganic product of fusion which has been cooled to a rigid condition without crystallization." Others suggest it is "a noncrystalline material obtained by a melt-quenching process"[11] or "an amphorous solid."[12] In this context we are not so concerned with what truly defines a glass, but rather what makes a glass moldable.

There really is no such scientific definition of a moldable glass. In some way or another almost any glass can be molded. The assignment of the word *moldable* to any glass is really a manufacturing term. The term is an assessment by the manufacturer of how readily the glass can be molded. Each manufacturer or molder will have its own definition of what is moldable, where the parameters are based on not only the glass material, but the mold materials, mold coatings, equipment used to mold the lens, and the process itself.

5.3.1 Moldable Glasses

There are many different types of glass for molding, provided by many different glass suppliers. Each supplier may have its own definition of what a moldable glass is, but for the most part *moldable* is defined by the glass transition temperature of the glass. For example, Schott defines its low T_g molding glass as glasses with T_g less than 550°C.[13] Figure 5.2 shows a variety of moldable glass prior to shaping into preforms.

All of the main glass manufacturers use prefixes before the designations for individual glass types, and most have special designations for glass intended to be used for precision glass molding. These designations typically mean the glass was specifically designed for use in precision glass molding. This does not mean that other glass types without the designation cannot be used. For instance, Ohara's PBH71 has been used prevalently in precision glass molding but does not have Ohara's L-designation, as it predates the naming practice. The L-designation for Ohara also indicates that the glass is environmentally friendly and PBH71 has a significant quantity of lead oxide. Schott also sells its N-type glasses in "P-Quality"[13] since these glasses meet Schott's requirements for glass molding though they were not designed expressly for molding. The naming conventions for the primary moldable glass suppliers for their primary lines of moldable glass are listed in Table 5.1. Contact information for each of these suppliers is listed at the end of this section.

The moldability of a glass is dependent on a number of factors, but for the most part, a moldable glass can be molded by most manufacturers. However, it is always important to confirm the manufacturability with your specific supplier. A simplified review of moldable glasses available from a number of suppliers reveals at least one hundred available glass types. A

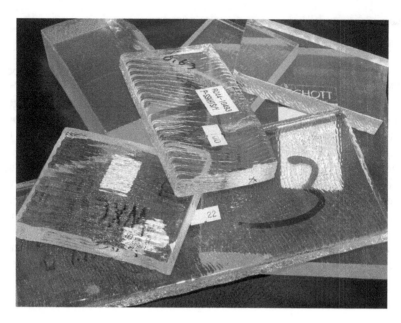

FIGURE 5.2
Moldable glass in plate form.

TABLE 5.1

Glass Manufacturer Designations for Moldable Glass Types

Manufacturer	Prefix	Description	Definition
CDGM	D	"Low softening point glass type without Lead, Arsenic"	$T_g < 618°C$[27]
Hikari	Q	N/A	$T_g < 607°C$[28]
Hoya	M	N/A	$T_g < 560°C$[29]
Ohara	L	"Low softening temperature glasses"	$T_g < 608°C$[30]
Schott	P	"Specifically developed Low T_g Glasses for precision molding"	$T_g < 550°C$[31]
Sumita	K	"Lead and Arsenic Free"	$T_g < 530°C$[32]

list is provided in Table 5.2. The primary molding attributes of these glasses that are typically listed by suppliers are included in the table. These are index of refraction (587.6 nm), n_d; Abbe number (587.6 nm), v_d; glass transition temperature, T_g; yield point, A_t; and coefficient of thermal expansion, α. Many other physical and optical attributes, such as transmission and dn/dt, should also be taken into consideration when selecting a glass; however, the attributes listed are the properties that are primarily of interest to the molding process.

TABLE 5.2

Selection of Moldable Glasses[a,b]

Code	Glass Designation	Manufacturer	ν_d	n_d	T_g (°C)	A_t (°C)	α^c $(10^{-7}/C)$
002206	K-PSFn2	Sumita	20.6	2.00170	480	514	73
020215	K-PSFn202	Sumita	21.5	2.01960	460	486	74
144178	K-PSFn214	Sumita	17.8	2.14352	425	449	83
434950	K-CaFK95	Sumita	95.0	1.43425	431	450	129
459900	K-PFK90	Sumita	90.0	1.45880	431	461	137
486845	N-FK51A	Schott	84.5	1.48656	464	490	148
486852	K-PFK85	Sumita	85.2	1.48563	452	484	129
488704	N-FK5	Schott	70.4	1.48749	466	557	100
497815	K-PFK80	Sumita	81.5	1.49700	461	483	134
497816	N-FK52A	Schott	81.6	1.49700	467	495	150
507705	K-PG325	Sumita	70.5	1.50670	288	317	143
512521	K-SKF6	Sumita	52.1	1.51200	432	516	95
516641	L-BSL 7	Ohara	64.1	1.51633	498	549	58
516641	D-K9	CDGM	64.1	1.51633	497	N/A	78
517642	D-K9L	CDGM	64.2	1.51680	492	N/A	69
518635	D-K59	CDGM	63.5	1.51760	490	N/A	75
518635	K-PBK40	Sumita	63.5	1.51760	501	549	54
523623	K-PBK50	Sumita	62.3	1.52250	481	518	67
525704	K-PMK30	Sumita	70.4	1.52500	528	572	83
527662	P-PK53	Schott	66.2	1.52690	383	418	160
528770	N-PK51	Schott	77.0	1.52855	487	517	141
543629	K-PG375	Sumita	62.9	1.54250	344	367	129
559625	L-PHL 2	Ohara	62.5	1.55880	381	407	99
565608	L-PHL 1	Ohara	60.8	1.56455	347	379	105
566610	K-PSK11	Sumita	61.0	1.56580	410	437	82
569713	K-GFK70	Sumita	71.3	1.56907	485	509	132
583594	L-BAL42	Ohara	59.4	1.58312	506	538	72
583594	D-ZK2	CDGM	59.4	1.58313	501	N/A	87
583595	M-BACD12	Hoya	59.5	1.58313	500	N/A	70
587596	P-SK57	Schott	59.6	1.58700	493	522	89
587596	D-ZK2L	CDGM	59.6	1.58700	509	N/A	71
587596	K-CSK120	Sumita	59.6	1.58700	498	536	72
589612	L-BAL35	Ohara	61.2	1.58913	527	567	66
589612	D-ZK3	CDGM	61.2	1.58913	511	N/A	76
589613	M-BACD5N	Hoya	61.3	1.58913	515	N/A	68
592607	D-ZK3L	CDGM	60.7	1.59170	506	N/A	77
592607	K-PSK100	Sumita	60.7	1.59170	390	415	95
592683	K-GFK68	Sumita	68.3	1.59240	512	536	129

(continued on next page)

TABLE 5.2 (continued)

Selection of Moldable Glasses[a,b]

Code	Glass Designation	Manufacturer	ν_d	n_d	T_g (°C)	A_t (°C)	α^c (10⁻⁷/C)
605503	ECO550	LightPath	50.3	1.60550	375	410	135
605504	CO550	Corning	50.4	1.60500	330	N/A	150
610579	D-ZK79	CDGM	57.9	1.61035	513	N/A	93
610579	K-VC79	Sumita	57.9	1.61035	516	553	72
613590	K-PSK200	Sumita	59.0	1.61305	386	412	100
632638	K-LaFK60	Sumita	63.8	1.63246	485	528	93
658369	K-PG395	Sumita	36.9	1.65800	363	392	124
669554	D-LaK70	CDGM	55.4	1.66910	531	N/A	77
670554	K-VC78	Sumita	55.4	1.66955	520	556	80
678549	L-LAL12	Ohara	54.9	1.67790	562	600	76
679549	D-LaK5	CDGM	54.9	1.67790	528	N/A	92
688313	P-SF8	Schott	31.3	1.68893	524	580	111
689311	L-TIM28	Ohara	31.1	1.68893	504	539	101
689311	D-ZF10	CDGM	31.1	1.68893	507	N/A	105
693337	K-CD45	Sumita	33.7	1.69320	470	507	91
*694531	D-LaK6	CDGM	53.1	1.69384	522	N/A	88
*694531	K-VC80	Sumita	53.1	1.69384	530	566	64
*694532	L-LAL13	Ohara	53.2	1.69350	534	575	76
*694532	M-LAC130	Hoya	53.2	1.69350	520	N/A	69
694563	K-LaFK55	Sumita	56.3	1.69400	514	556	73
714389	D-ZBaF38	CDGM	38.9	1.71430	471	N/A	104
714389	K-ZnSF8	Sumita	38.9	1.71430	518	546	49
723292	D-ZF20	CDGM	29.2	1.72250	490	N/A	127
723292	K-CD120	Sumita	29.2	1.72250	508	549	92
731405	L-LAM69	Ohara	40.5	1.73077	497	529	86
731405	M-LAF81	Hoya	40.5	1.73077	500	N/A	89
731405	D-LaF79	CDGM	40.5	1.73077	618	N/A	93
733489	L-LAM72	Ohara	48.9	1.73310	565	608	66
735488	D-LaF82L	CDGM	48.8	1.73485	558	N/A	67
743493	L-LAM60	Ohara	49.3	1.74319	541	581	74
743493	M-NBF1	Hoya	49.3	1.74330	555	N/A	54
743493	D-LaF53	CDGM	49.3	1.74330	549	N/A	72
756456	K-VC82	Sumita	45.6	1.75550	526	563	59
772500	K-LaFK50	Sumita	50.0	1.77200	560	592	73
806404	L-LAH81	Ohara	40.4	1.80610	566	602	58
806407	M-NBFD130	Hoya	40.7	1.80610	560	N/A	N/A
806409	L-LAH53	Ohara	40.9	1.80609	574	607	59
806409	P-LASF47	Schott	40.9	1.80610	530	580	73
806409	D-ZLaF52A	CDGM	40.9	1.80610	538	N/A	78
809404	L-LAH84	Ohara	40.4	1.80860	527	568	64

TABLE 5.2 (continued)

Selection of Moldable Glasses[a,b]

Code	Glass Designation	Manufacturer	ν_d	n_d	T_g (°C)	A_t (°C)	α^c (10⁻⁷/C)
810409	D-ZLaF52LA	CDGM	40.9	1.81000	546	N/A	76
810410	K-VC89	Sumita	41.0	1.81000	528	559	64
814370	D-ZLaF82	CDGM	37.0	1.81474	550	N/A	67
815370	M-NBFD82	Hoya	37.0	1.81474	550	N/A	59
839239	K-PSFn3	Sumita	23.9	1.83917	477	515	93
844248	K-PSFn4	Sumita	24.8	1.84400	469	510	103
847238	L-TIH53	Ohara	23.8	1.84666	561	598	78
851416	K-VC99	Sumita	41.6	1.85060	616	653	60
853390	K-VC90	Sumita	39.0	1.85280	583	633	77
854404	L-LAH85	Ohara	40.4	1.85400	614	659	65
854406	D-ZLaF85L	CDGM	40.6	1.85370	612	N/A	76
864406	L-LAH83	Ohara	40.6	1.86400	608	658	68
887350	K-VC91	Sumita	35.0	1.88660	589	638	77
902251	L-NBH54	Ohara	25.1	1.90200	547	586	81
903310	L-LAH86	Ohara	31.0	1.90270	578	610	61
906214	P-SF67	Schott	21.4	1.90680	539	583	74
907212	K-PSFn1	Sumita	21.2	1.90680	498	543	74
921224	K-PSFn5	Sumita	22.4	1.92110	463	495	108

[a] Data are provided for reference only and are not intended for use; the individual suppliers should be contacted for official data.

[b] Data references: CDGM,[27] Corning,[32] Hoya,[28] LightPath,[33] Ohara,[34] Schott,[30] and Sumita.[31]

[c] Coefficient of thermal expansion, α, data are provided for the reference temperature range of ~ −30 to ~ +70°C.

From the table we can see the optical designer has a wide range of selections; the variable ranges are

$$17.8 < \nu_d < 95.0$$

$$1.434250 < n_d < 2.143520$$

$$288.0 < T_g < 618.0$$

$$49 < \alpha < 160$$

The data from Table 5.2 are plotted in Figure 5.3, which shows the significant variety available. Figure 5.4 shows a revised selection of glasses identifying the manufacturer and eliminating duplicate glass codes, showing that even without manufacturer redundancy there are a significant number of options.

These charts are very helpful but do not present a complete picture; without critical information such as melt frequency, the frequency at which a glass is manufactured, a designer may request a very expensive design with very long lead times. Moldable glasses are no different, the availability and cost of glass

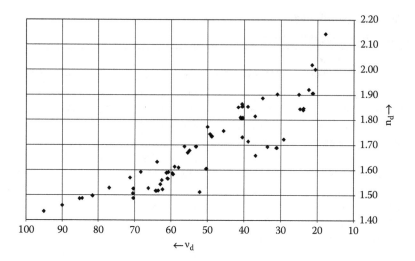

FIGURE 5.3
Selection of moldable glasses.

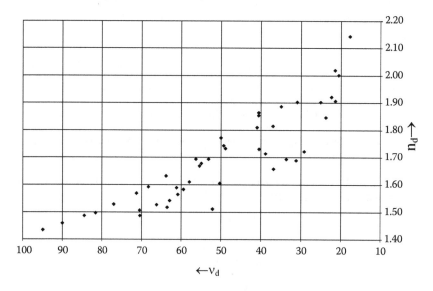

FIGURE 5.4
Selection of moldable glasses with material grade.

should always be taken into consideration. Glass producers enjoy economies of scale just as lens manufacturers do, so despite the fact there are hundreds of glasses to choose from, most suppliers will narrow their focus to several primary grades. For example, Ohara specifically recommends L-BAL42 as the "preferred material of choice for high volume mold pressing production."[22]

It is also important to note that the same basic glass may be available from a number of suppliers. For example, CDGM's D-Lak6, Sumita's K-VC80, Ohara's L-LAL13, and Hoya's M-LAC130 are very similar to one another. The properties of these glasses are shown in Table 5.2 and indicated by the asterisks. For most applications, these glasses would be readily replaceable with the other manufacturers' materials. There are, however, subtle differences, and it is important to do a detailed review prior to making any glass substitutions in a design.

A special type of glass used in the molding of lenses for use in short- to long-range infrared wavelengths is chalcogenide glass. These glasses will be reviewed in greater detail in the next chapter.

5.3.2 Glass Properties

The single most discussed attribute for a moldable glass is its glass transition temperature or glass transformation temperature, T_g. The T_g of a glass is determined by measuring its thermal expansion, typically by using a dilatometer. Figure 5.5 shows the typical shape of a thermal expansion curve. Note that there are two distinct regions on the curve prior to reaching the yielding temperature. Once the glass is heated beyond a certain temperature, it begins to expand much faster. The intersection between the two expansion rates is considered the T_g, as shown in Figure 5.5. The rates of expansion, however, are based on the rate of heating or cooling. This is important in glass molding because it indicates the T_g is dependent on the cooling/heating rate of

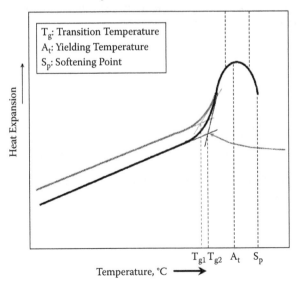

FIGURE 5.5
Determination of glass transition temperature, T_g.

the molding process, T_{g1} vs. T_{g2}, in Figure 5.5. What this means for precision glass molding is that the molding process itself can have an impact on the final optical properties of the molded lens, specifically a change in the refractive index. This will be discussed in Section 5.7.

5.3.3 Environmental Concerns

In recent years, there has been a significant effort for manufacturers to become more environmentally conscious. Many manufacturers now require their suppliers to reduce or eliminate the use of hazardous materials in their manufacturing processes. There are also new regulations for controlling these substances. The manufacture of glass is no exception to these changes.

Optical glass manufacture has historically used metal oxides and heavy metal refining agents. Among those used were lead oxide (PbO), arsenic trioxide (As_2O_3), and cadmium. Most manufacturers have eliminated these materials from the manufacture of optical glass, while some maintain some limited supply for special applications. Hoya, for example, has completely eliminated cadmium from its catalog.[14] The inclusion of lead oxide was especially advantageous for precision glass molding, as PbO in glass lowers the melting point and increases the index of refraction. However, it is estimated that the global manufacture of leaded glass may release up to 2 to 5 kg/ton of lead emissions to the atmosphere.[15]

One of the leading regulations for limiting the use of hazardous materials is the European Restrictions on Hazardous Substances (RoHS). In order to be RoHS compliant, the homogenous material content of a product must not exceed the following concentration values:

Lead: 0.1% by weight

Mercury: 0.1% by weight

Cadmium: 0.01% by weight

Cr^{6+}: 0.1% by weight

Polybrominated biphenyls (PBBs): 0.1% by weight

Polybrominated diphenyl ethers (PBDEs): 0.1% by weight

Homogenous material means a unit that cannot be mechanically disjointed into different materials.

Many companies now require RoHS compliance, and therefore there has been a significant move away from glasses containing any of these materials.

5.3.4 Glass Manufacturers

The following is an alphabetical list of glass suppliers that promote environmentally friendly, low T_g glasses for precision molding. Information is provided for their location closest to the eastern United States; further information can be obtained by contacting them directly:

CDGM Glass Co. Ltd.
CDGM Glass Company USA
3495 Winton Place, Building D, Suite 2
Rochester, NY 14623
Phone: 585-427-9184
Website: www.cdgmglass.com, www.cdgmgd.com

Hikari Glass Corporation (Nikon)
Hikari Glass Corporation
226 Vailco Lane
Lakeway, TX 78738
Phone: 512-608-0070
Website: www.hikari-g.co.jp

Hoya Corporation
Hoya Corporation USA
Optics Division
3285 Scott Boulevard
Santa Clara, CA 95054
Phone: 800-818-HOYA
Website: www.hoyaoptics.com, http://www.hoya.co.jp

Ohara Corporation
Ohara Corporation
50 Columbia Road
Branchburg, NJ 08876-3519
Phone: 908-218-0100
Website: http://www.oharacorp.com/

Schott North America, Inc.
Schott North American, Inc.
Advanced Optics
400 York Avenue
Duryea, PA 18642
Phone: 570-457-7485
Website: http://www.us.schott.com/

Sumita Optical Glass, Inc.

Sumita Optical Glass, Inc.

4-7-25 Harigaya, Urawa-Ku, Saitama-City

Saitama, 330-8565 Japan

Phone: +81-48-832-3165

Web site: http://www.sumita-opt.co.jp

5.4 Glass Preforms

The first step in a glass molding system starts with a glass preform. As described, this is simply a preformed piece of glass from which the lens will be manufactured. It is important to have a thorough understanding of the preform manufacturing process, as the quality of the glass preform has a direct impact on the final optic. A preform may take many different shapes and sizes, each with its own distinct advantages and disadvantages.

5.4.1 Spherical or Ball Preforms

The most prevalent and well-known preform for glass molding is the spherical or ball preform. Figure 5.6 shows a variety of ball preform sizes and materials. The spherical preform has many advantages. The manufacturing process is a mature technology that is also used to manufacture ball lenses. In fact, there is basically no difference between most spherical preforms and a ball lens.

Ball preforms are manufactured through a custom grind and polishing process. The material will start out in a slab (sometimes even a gob) form and is then cut to small sections approximating the size of the ball. These cubes are then tumbled to remove sharp edges. These ball-like structures are then rough ground to the near final diameter. The rough-ground balls are then finish ground and followed up with a final lapping or polishing procedure. During these processes a number of failure modes can occur that can impact the final molding of the lens. Typical issues are cracks, scratches, and haze. These defects will be directly transferred to the final lens if not removed from production. A high-quality preform is required to mold a high-quality lens. The molding process is essentially a reshaping rather than a transitional process. Any defect in the preform will be passed to the lens, regardless of the preform shape.

The manufacture of a ball preform requires process control of only one physical dimension, the diameter of the ball. This single physical parameter drives the volume, and therefore the mass, of the ball. The dimension and maintenance of the spherical shape of the ball are the controlling dimensions.

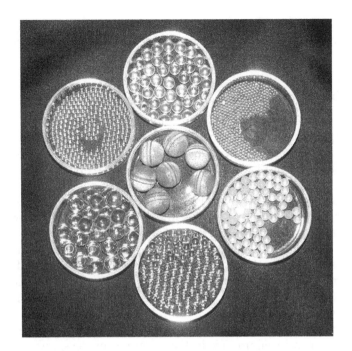

FIGURE 5.6
Selection of ball preforms in various states of manufacture.

A ball lens therefore has few controlling parameters, specifically, material, diameter, tolerance on diameter, and surface quality. The surface quality is usually controlled by a cosmetic inspection for scratch/dig criteria and opacity. For these reasons a ball preform is typically the most accurate of all preforms for precision glass molding.

Although the quality of a ball preform depends on many things, the type of glass and quality of the supplier are first and foremost. This is not a discussion of ball lens manufacturing, but to generalize, the softer the glass, the harder it is to polish, the more difficult it is to handle, and the more prone the preform is to scratch and dig type defects. Harder glass tends to be lower in cost and easier to manufacture but usually has a higher T_g.

Many ball preform manufacturers are limited on size, but most can conveniently manufacture preforms in the range of 1 to 8 mm in diameter. Smaller or larger preforms will require manufacturers with specialized equipment or processes and may be limited in capacity. Very small preforms, less than 0.75 mm in diameter, are seldom seen and are very difficult to manufacture. Whereas balls are easily manufactured on a custom basis, cost and time to market will almost always be improved by using a readily available size.

The disadvantage of using ball preforms is that you are basically making a lens to make a lens. This correlates to a higher cost. However, economies of scale, machines designed for volume manufacture, and the simplicity of operation and inspection have greatly reduced the pricing of ball preforms. In some cases the pricing of ball preforms can even be comparable to gob preforms.

5.4.2 Gob Preforms

The other prevalent preform is the gob preform. A gob is simply a prescribed volume of glass used for pressing. The shape of gob preforms will vary from manufacturer to manufacturer. Ohara describes its gobs as "Oval-convex,"[30] whereas Sumita will provide convex-convex, convex-plano, or convex-concave gobs; hence, the nebulous term *gob* is a very good description. Some very ball-like gobs are shown in Figure 5.7. These gobs are so close to a ball shape that it is difficult to discern them from a ball with the naked eye. The gob was developed to minimize the preform manufacturing cost. Gob manufacturing equipment is highly proprietary, and there are no known commercially available solutions. This is largely because, unlike ball preforms where there is a significant market for ball lenses, there is little value outside of glass molding for preformed gobs. Gobs may even sometimes be used as the precursor to a ball preform.

The primary method of manufacturing gobs is dropping. The dropping method is shown in Figure 5.8. In the dropping process, gobs are formed from continuously flowing molten glass. The glass is heated in a crucible and discharged through a nozzle at the bottom. Heat is applied and maintained at both the crucible and the nozzle. The nozzle is designed specifically for the size of preform required. Therefore, custom preforms require custom nozzles and dedicated runs. The process relies on the volume and viscosity of the glass to continuously feed the glass; additional glass must be added to the crucible for consistent flow. The design of the system and the nozzle has a significant impact on the final shape of the gob. Different systems and

FIGURE 5.7
Gob preforms.

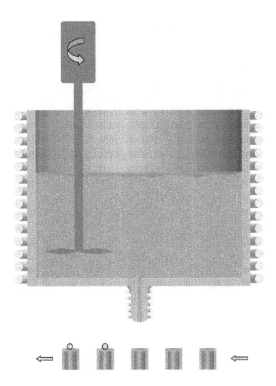

FIGURE 5.8
Dropping method for gob preform manufacture.

different nozzles will lead to shape variations. Once the gob is discharged it is "caught" on specialized plates indexed by a rotational conveyor. The speed of the conveyor must be matched with the speed of the process. Poor gob formation can lead to poor quality surface finishes, striae, and even cracks within the gob. These defects will be subsequently passed on to the pressed lens. Gobs are intended for high-quantity applications and are specified based on volume. Gobs also are more restrictive on size and material. Many suppliers only supply a limited number of glass types in gob form, and variations in tolerances are large, as they are very dependent on size.

5.4.3 Other Preforms

Almost any other shape of glass could be used as a preform, but the most common are driven by cost or complexity of the shape of the final lens. Figure 5.9 shows a number of different preform shapes. Flat (plano-plano) or disc preforms are commonly used as low-cost, material-efficient shaped performs, but typically require a vacuum molding process to eliminate trapped gas, as discussed later in this chapter. Disc preforms can be polished to a very high quality and with a very high accuracy on thickness, which can

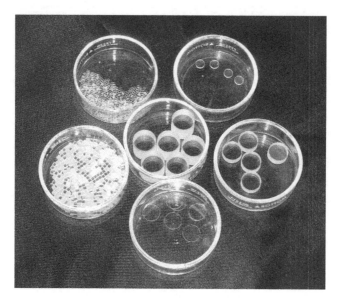

FIGURE 5.9
Selection of custom preforms.

lead to a high-quality end product. These preforms are used for V-groove arrays, wafer type arrays, and many thin molded components. For expensive base materials they are advantageous due to the material efficiency of manufacturing the preform, as there is little waste when compared to ball preforms or near-net shape preforms. Disc preforms are also the preferred embodiment for negative lenses.

Near-net shape preforms are a close approximation to the final product prior to the molding process. These preforms will typically require little change in shape during the molding process, but are the most expensive of all preforms to manufacture. They are advantageous for complex geometries and for larger shapes.

Recently, Schott has disclosed a process for using precision-rod-based preforms for manufacturing linear arrays of lenses.[16] The options for preforms are limited only by the availability of the glass shapes desired—the challenge is to determine which shape makes the most sense for the specific application.

5.4.4 Preform Selection

Preform selection is the responsibility of the glass molder. The molder will consider many factors when selecting a preform shape: material, shape and size of finished product, volume of order, and required quality. Some lenses or components may have a clear solution such as a cell phone lens, which would clearly use a gob or ball preform, whereas a large, low-volume

TABLE 5.3

Comparison of Preform Shapes

Preform Shape	Relative Cost	Quality	Typical Size
Ball	$	Very high	$1 < \varnothing < 8$ mm
Gob	$	High	0.1–0.7 cc
Disc	$	High	Any
Near net	$$$	High	Any

lens would use a disc or near-net shape. Table 5.3 summarizes the relative performance of a number of preform shapes.

5.5 The Precision Glass Molding Process

Precision glass molding is a rather straightforward process. It is essentially a high-temperature compression molding process within a controlled environment. The process starts with the glass preform. The glass preform is placed in the glass molding machine. The preform may or may not be preheated to a specific temperature prior to insertion into the machine. Preform insertion may be a manual or automated process. The molds may be a single set or have a number of individual cavity sets. Lens or component-specific molds are already installed in the glass molding press. The molds may also be preheated prior to insertion of the preform. Once the preform is inserted into the press, the molding chamber is then evacuated with an inert gas. Typically the chamber is evacuated with nitrogen. The heating process begins next. The preform will begin to ramp to a temperature set point. This first temperature set point is dependent on the T_g of the glass, in order to get the glass to a specific viscosity range, approximately 10^9 dPa·s. The preform will then be compressed between the molds. The amount of force applied to the preform is controlled throughout the process. Temperature is held stable for a set period of time. The lens is then allowed to cool, during which time the load continues to be applied until the temperature of the glass drops beneath the strain point of the glass. Once this point is reached, the load is removed and a more rapid rate of cooling is typically applied in order to improve throughput. The finished product is then removed from the molds and then further cooled. A typical processing cycle is shown in Figure 5.10. Additional annealing operations may also be applied after molding. This is a simplification of the process, and there can be many variations based on the manufacturer, equipment, and glass types.

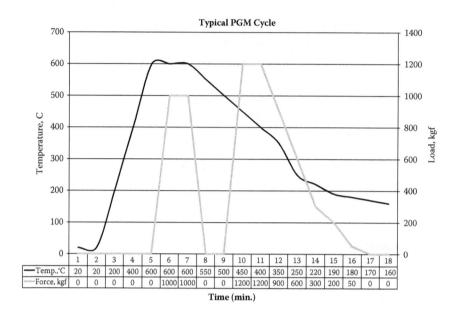

FIGURE 5.10
Typical precision glass molding process.

5.5.1 Press Molding Equipment

There are a number of precision glass molding presses commercially available, although many companies manufacture their own molding equipment.[17] Toshiba Machine's GMP series is the most well-known commercial equipment and has the longest documented history. The construction of a glass molding press is rather simple, especially when compared to an injection molding machine. A typical molding press consists of five primary components or systems: a heating system, a gas transfer system, a load transfer system, a control system, and component-specific tooling (Figure 5.11). The challenge of these presses is not so much in the basic principles, but rather the environment in which the process must take place, yet another reason for the pursuit of lower T_g glasses. The heating system must be able to heat the glass beyond its T_g. This is normally accomplished with an infrared heating system. The gas transfer system is used for the introduction of the inert gas to remove oxygen and to provide a means of forced convection. The load transfer system generates the force needed to shape the preform into its final form. The component-specific tooling consists primarily of the molds and other components required to form the shape of the lens. The control system consists of the electronics and software to manage all of these actions.

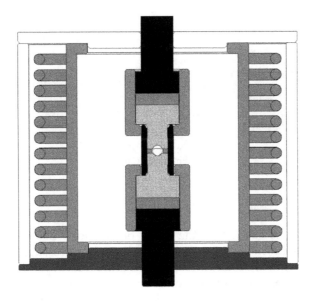

FIGURE 5.11
Typical precision glass molding press.

5.5.1.1 Transfer Molding

A special type of machine has been developed for high-volume small lenses.[18] These machines are referred to as glass molding transfer presses. Transfer molding machines operate on the same principles as the standard molding concept but are designed to accelerate the delivery or cycle time between lenses. The machines accomplish this, not by speeding up the process, but by creating many "stations" that transfer from step to step. This is accomplished by having many stations or mold sets that move through the molding cycle. The preforms are loaded at the first station; the mold set with a preform then moves to a heating phase, while another mold set enters the loading station. The mold set in the heating phases then moves to a load station, and the process continues throughout the mold cycle discussed previously. The typically discrete process of precision glass molding is turned into an approximation of a continuous flow process, and lenses can be manufactured at a much faster rate. This method obviously requires a significant investment in tooling for a large number of cavities or mold stations. Molds can be removed or introduced to the process without stopping the machine. Transfer molding is primarily used for large-volume applications such as cell phone lenses.

5.5.1.2 Vacuum Molding

For specific preform-mold combinations, the high quality of the surface of the preform and the high quality of the mold surface can lead to sealing

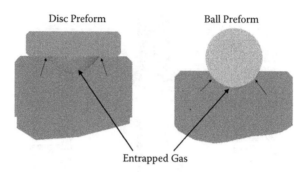

FIGURE 5.12
Gas entrapment in precision glass molding.

of oxygen or inert gas within the cavity of the mold. Gas entrapment can lead to defects in the lens itself or more difficult processing of the lens in order to obtain the final shape. This combination is most prevalent with disc preforms and concave mold cavities. It can also become an issue for concave molds with large included angles and ball or gob preforms (Figure 5.12). In this case it becomes desirable to mold under vacuum conditions or modify the process and design of tooling to allow for the alleviation of the gas. Molding equipment may introduce a vacuum after the purging process, which can reduce the issue. It is best to avoid design conditions that can lead to entrapped gas if at all possible, as they lead to complexity, and therefore cost, in the molding process.

5.5.2 Molding Classifications

There are a number of ways to classify glass molding processes. The following method differentiates based on the mold material required to mold the glass, and hence the T_g of the glass itself. This method includes three classifications: ultra-Low T_g glass molding, low T_g glass molding, and high T_g glass molding. Each method is discussed in detail below, and a simplified comparison of the different types of molding processes is shown in Table 5.4.

TABLE 5.4

Mold Process Classifications

Process	Ultra-Low T_g	Low T_g	High T_g
Glass T_g	$T_g < 400°C$	$T_g < 600°C$	$T_g > 600°C$
Molds	Electroless nickel-phosphor over base material	Carbides or ceramics	Carbides or ceramics
Mold manufacturing process	Single-point diamond turning	Micro-grinding	Micro-grinding
Cost of tooling	Low	High	Very high
Tooling life	Low	Mid	Very low

The table describes the main differences between the processes and summarizes the process that could be used for individual glass types. It is important to note that this is a generalization; a glass with a T_g of 380°C could still be molded using the low T_g process. The equipment used to press the lens is basically the same; the primary difference is the manufacturing method of the molds.

5.5.2.1 Ultra-Low T_g Glass Molding

The ultra-low T_g glass molding process is seldom used. This process was developed many years ago and was initially based on lead oxide glasses due to the advantages of high refractive index with low T_g. The use of lead-oxide-based glass resulted in a number of glasses with T_gs less than 400°C, such as Corning's CO-550, T_g = 330°C (~25% PbO); Ohara's PBH-56, T_g = 386°C (>70% PbO); and Ohara's PBH-71, T_g = 389°C (~80% PbO). The implementation of RoHS restrictions on lead in glass has driven many glass suppliers to discontinue their lead-based glasses. One of the primary advantages of ultra-low T_g glass molding when it was developed was that single-point diamond turning (SPDT) could be used to manufacture molds rather than precision micro-grinding. Developments in grinding technology at the time were rather limited, requiring custom-developed equipment with significant internal know-how in order to be able to manufacture molds of comparable quality to SPDT. There was little knowledge base or experience with grinding of optical surfaces for carbide or ceramic molds.

The ultra-low T_g glass mold manufacturing process is very similar to the traditional manufacture of molds for injection molded plastic optics. The process starts with a raw material that has suitable characteristics for the temperatures of processing. It is desirable to pick a material that has a coefficient of thermal expansion close to those of both the glass to be pressed and the plating material. Once the raw material is selected, it is then machined to a near-net final shape. This shape is smaller than the final mold to allow for the plating process. The mold is then plated. The plating is an optical quality electroless nickel-phosphor intended for SPDT. Once plated, the mold is then single-point diamond turned. The design of the mold will compensate for the expected changes in the surface due to the processing of the lenses. It is very rare for molds manufactured in this manner to be polished after diamond turning. The plating has an affinity for smearing rather than polishing, making it risky and very difficult to improve the surface roughness of the molds. Once turned, a hard coating is applied to the mold to extend mold life during processing. Once coated and inspected, the mold is ready for making lenses. The molds at the various stages are shown in Figure 5.13. The limitation on T_g is driven by the breakdown of the electroless nickel-phosphor plating, which occurs around 400°C, depending on the specific type of plating.

The disadvantage of this process is that the materials are inherently softer, leading to shorter mold life and higher costs for larger volumes. The reduced

FIGURE 5.13
Molds in the various manufacturing stages of the ultra-low T_g mold manufacturing process.

mold life can be partially offset by the increased speed and quality at which these molds can be manufactured. The lower cost of tooling, when compared to other methods, makes this process a viable solution for manufacture of smaller volumes.

5.5.2.2 Low T_g Glass Molding

The low T_g process is the prevalent method used today. The process follows steps very similar to those outlined above, but is based on much harder substrates. The material substrate for the low T_g glass molding process is typically a grade of silicon carbide (SiC), tungsten carbide (WC), or other material, such as Si_3N_4.

The mold manufacture process starts with the premachining of a blank in order to create a premold, a piece of raw material that has the basic shape already ground in. This is done to facilitate faster turnaround of the final mold geometry. This step could easily be skipped and the premold manufactured directly from raw material. A premold is simply a near-net shape that is ready for final grinding. A premold is manufactured using a number of traditional grinding techniques in order to reduce cost and minimize the amount of work during the final grinding phase. Unlike the plating process described above, the premold requires excess material for the final precision grinding process to remove. The final grinding process is then used to create the optical surface. The grinding process is done on the same basic machine as used for SPDT. These machines, however, have been reconfigured for the precision grinding process. High-speed air-bearing, air turbine spindles are installed to drive the grinding wheels. Mechanisms for the adjustment of spindle height are added. A rotary indexing table is usually added as well, to allow 45° grinding, along with coolant systems and shrouding. Grinding wheel dressing stations may be included to improve productivity. Two primary methods are used for grinding carbide molds, wheel normal and 45° grinding. Wheel normal grinding is used for concave molds and shallow convex molds. Steeper concave molds require 45° grinding, a more difficult, time-consuming, and expensive process. These processes are now called deterministic micro-grinding, with the emphasis on *deterministic*. *Deterministic* describes a system whose time evolution can be predicted exactly. This is a bit of a stretch for the process of grinding optical molds; however, the process has come along significantly in the past decade, making precision-ground carbide molds a reality. Form errors of 0.199 μm (PV) with surface roughness (R_a) less than 4 nm in WC (0.5% wt Co) have been reported.[19]

Once the mold is ground, it may be subjected to a postpolishing step in order to remove some of the residual grinding marks and to improve the surface roughness. This polishing may come at the expense of degraded form error and should be evaluated on a case-by-case basis. Once the form is acceptable, a hard coating is applied to the mold.

5.5.2.3 High T_g Glass Molding

High T_g glass molding can be defined as molding glasses with T_g greater than 620°C. Many glasses with T_g greater than 620°C can be easily replaced with a lower T_g glass, which would clearly be advantageous. High T_g, therefore, tends to refer to very high T_g glass with specific properties that are not easily reproducible with other glass types. The main focus is on materials such as quartz glass, T_g = ~1,200°C. High T_g molding is basically the same process described above but conducted at much higher temperatures. The design and manufacture process becomes more difficult due to the much higher processing temperatures, and material selection becomes essential. Many of the same materials are used as discussed above, but it has been stated that amorphous carbon has had the most success with some mold processes.[25] There are molding machines capable of handling these much higher temperatures, some able to heat molds up to 1,500°C.[20] These higher temperatures simply exasperate the issues associated with precision glass molding. Oxidation, coating interactions, and thermal mismatch issues are all much more difficult to resolve. The result is significantly reduced mold life and much higher costs to manufacture.

5.5.2.4 Other Molding Processes

Other molding type processes that may not easily fit into the above categories are of interest because the end result is similar. These processes include the direct sintering of zinc sulfide (ZnS) lenses and the automated casting of chalcogenide glasses. Both processes result in the creation of a precision molded lens.

After many years of manufacturing lenses from sintered ZnS,[21,22] a powder metallurgy process has been developed in which ZnS is directly sintered and molded into a finished lens shape.[23] This process is very similar to precision glass molding, with the exception that the preform is replaced with constituent powders. The powder is introduced to a set of molds, and then a sintering and forming process takes place, resulting in a final lens shape. A similar process has been developed for chalcogenide glasses. In 2007, a novel approach for the automated casting of chalcogenide glasses was patented.[24,25] This process is better defined as a casting process rather than a molding process due to the liquid nature of the glass at the time of introduction to the molds. The process is very similar to the transfer molding process described previously, with the exception of the material delivery stage. The process begins with empty mold sets. The mold sets are then preheated and moved to a casting chamber. In the casting chamber liquid glass is introduced to the individual mold sets. This stage is then followed by pressing and cooling cycles. Similar to transfer molding, the process can significantly improve capacity and cycle times over traditional glass molding, with the requisite significant investment in a large number of molds sets. Both of these processes are similar in that the material is formed at the time of process. The advantage is a very efficient use of material, but with an increase in complexity of processing and operation of equipment.

5.5.3 Mold Design

Regardless of the process used to manufacture the mold, it must be designed to replicate the final optical prescription of the lens. This is not as simple as using the same equation as the lens surface prescription because the mold must compensate for thermal expansion and shrinkage. In order to minimize the effect of thermal shrinkage, the mold must be uniformly heated. Nonuniform heating may require additional design compensations. There are many techniques for the design of these surfaces, from empirical studies[31] to finite element analysis.[26]

5.5.4 Mold Coating

Molds are almost always coated prior to use. Due to the extreme nature of the process, it is important to limit the interaction of the glass with the surface of the molds. Simplistically, this is the purpose of coating the mold, to limit the interactions in order to extend the life of the mold, and to thereby reduce the cost of the end product. The precision glass molding process is very caustic, and the molds are subjected to numerous thermal cycles and mechanical interactions. The manufacture of molds is an expensive process; the mold coating protects that investment.

There are numerous failure modes associated with the lifetime of a mold for precision glass molding. The primary failure modes are associated with damage or localized delamination of the mold coating, to which there are a number of contributing factors. Oxidation of the mold generates oxides along the surface of the molds. These oxides can be considered unwanted contaminates in the molding process that will lead to premature failure of the mold. The contamination can damage the mold coating or the glass itself. Oxidation can be reduced in a number of ways, such as limiting the oxygen content during the press cycle; this is accomplished by introducing an inert gas, e.g., nitrogen, or a vacuum during the pressing process. This is primarily a condition controlled by the molding process and the equipment used to press the lens. Material selection is also important, as lower T_g will result in lower processing temperatures, and therefore lower likelihood of the material to produce oxides. The higher the temperature for pressing the glass, the more likely it is that the glass will stick to the mold surface.[17] Many carbide manufacturers now promote their materials based on resistance to oxidation. This is typically reported as the percentage of weight change due to the formation of the oxides. Material wear will also lead to the propensity of the material to have particles or oxides break loose and enter into the mold process, thereby leading to premature failure of the molds. The chemical interaction between the glass and coating or substrate can lead to the loss of transparency of the glass itself.

Mold coatings fit into a number of categories: single-layer carbides, nitrides such as TiAlN, CrN, TiBCN, or TiBC, multilayer carbides and nitrides, diamond-like coatings, and precious-metal-based alloys such as PtIr. Only the careful selection of mold coating, molding material, molding process, and glass will lead to the most cost-effective mold solution. As companies vary in their selection of any one of these parameters, there can be a significant difference in their performance. The longer the life of the mold, the smaller the number of molds required over a product's lifetime. The cost of the mold is typically amortized into the cost of the lens, as they are a direct cost. The longer the life of the mold, the lower the cost of the lens produced from it. Hence, a significant amount of time and effort has been put forward in the development of extending mold life.

5.5.5 Process Classifications

The precision glass molding process can be further classified into whether or not the final molded shape is volumetrically controlled.

5.5.5.1 *Volumetric Molding*

Volumetric molding is the molding of a glass into a predefined shape in which it is physically constrained in all directions. Volumetric molding requires tightly controlled preforms, since the volume of the preform must closely match the space it will fill. The tooling for volumetric molding must also generate the outside diameter of the lenses. Greater load capability of the pressing machine is also required in order to get the glass to fill the mold. This approach will result in a finished lens at the end of pressing. The edges will be somewhat rounded and are normally not edged, as there is no need. These edges are where the slight variations in volume are typically observed. The primary advantages of this method are that postprocessing is not required and the optical centering of the lens is controlled very accurately.

5.5.5.2 Nonvolumetric Molding

The molding of a lens into a shape in which the volume is not controlled and the shape is allowed to settle may be called nonvolumetric molding or free molding. The advantage of free molding is that the preform does not need to be specific to the design; many different preform sizes could be used to make the same lens. Free molding requires less force to mold than volumetric molding, but increases the difficulty in controlling center thickness. A free molded lens requires postprocessing to get the lens to its final shape. Postprocessing relies on the accuracy of the edging machine for the accuracy of centration of the optical surfaces to the outside diameter of the lens. Nonvolumetric molding is also the only option for extremely small lenses that are beneath the capability of preform manufacturers.

5.5.6 Insert Molding

Similar to injection molding, insert molding can also be accomplished in the precision glass molding process. Inserts are normally a stainless steel alloy; 304L, 416, and SF20T are typical grades. SF20T may contain lead (0.1 to 0.3%) and should be reviewed against any environmental restrictions. It is essential for the coefficient of thermal expansion (CTE) of the insert to be close to that of the glass in order to reduce the residual stress in the lens after molding. Insert molding can aid in assembly of the lens in the end product and will normally result in a lens that is centered in its holder much more accurately than if it was bonded. The insert molding process can create a hermetic seal between the glass and metal ring. In addition, it is common to see gold plating on inserts for improved laser welding. While housings may discolor due to the high temperatures of precision glass molding, this is purely cosmetic. Figure 5.14 shows some examples of insert molded lenses.

5.6 Postprocessing

The flexibility of precision molded optics is greatly enhanced by the addition of postprocessing. Postprocessing allows the creation of shapes that may not be readily achievable by direct molding. These changes may be to reduce size, reduce weight, add mounting features, or improve stray light performance. There are a number of standard postprocessing techniques. One of

FIGURE 5.14
Examples of insert molded lenses.

the simplest postprocessing procedures is traditional edging and grinding. Others may include nontraditional grinding operations, postpolishing, or dicing. Figure 5.15 shows a number of examples of postprocessed precision glass molded lenses.

5.6.1 Centering and Edging

Depending on the production method of the molded optic, the lens may actually require edging and centering. A free molded lens requires grinding to generate the finished outside diameter; a ground lens then obviously requires edging to relieve the sharp edge condition. As discussed earlier, a lens pressed without a mold cavity will press to an uneven shape. These lenses clearly require some postprocessing. The lenses are centered and edged using either inexpensive traditional or much more advanced fully automated machines.

Traditional centering of a lens relies on the spherical nature of a lens to self-center on the machine; this is much more difficult to achieve for aspheric surfaces. The end result is it is much more difficult to achieve tight tolerances on optical centering for aspheric surfaces on all but the most advanced machines. In situations where tight centration tolerances are required, a volumetric molding process, which inherently has excellent centration, should be used.

FIGURE 5.15
Examples of postprocess PGM lenses.

5.6.2 Dicing and Nontraditional Grinding

Many product designs are driven by size; the more compact a design, the more successful it is expected to be in the market. Therefore, it may be advantageous to consider additional postprocessing. While some shapes may be directly molded, more aggressive shapes can be created by using a dicing process or even nontraditional grinding. While dicing is limited to linear cuts, it can be a highly cost-effective and efficient method for reshaping a lens. Advanced grinding and centering machines can generate nontraditional shapes, step features, and even measure center thickness *in situ*. The advantage gained with postprocessing must be balanced with the costs associated with these operations. While a loosely toleranced single dicing operation would be relatively inexpensive in high volume, a tight-tolerance multiple-step process will have a significant impact on cost for a low-volume run.

Dicing of an optical component is a frequently used option. This is accomplished with wafer dicing saws using custom blades, arbors, and fixtures, but in volume is not overly expensive. These lenses would be ganged on the wafer saw, thereby increasing throughput and lowering cost, an inexpensive solution for volume needs. Dicing will result in some edge chipping, normally on the order of 50 to 100 μm, and tolerances can be held extremely tight if required, to less than 5 μm.

Postprocess by dicing is also advantageous when making small lenses in volume. A lens array can be made and then diced into individual lenses. Very small lenses that would be virtually impossible to mold in the traditional sense can be manufactured in this manner. The per lens cost can be very low in high volume once the high cost of the array tooling is overcome.

FIGURE 5.16
Example of integral molded features.

5.6.3 Postprocessing versus Integral Molding

Injection molded plastic optics have much more freedom for integrally molded features than precision glass molding. Almost all glass molded lenses will have flanges that are advantageous for mounting of the lens. Additional features and shapes can be molded but may be limited. Square-shaped lenses can be molded for specific designs only, as the preform must be accurately fit to the design. The example shown in Figure 5.16 includes a notched corner as an alignment feature to orient the lens during assembly. V-grooves and other shapes can also be molded in glass.

5.7 Design

The section on design was left to the end of the chapter because it is always imperative in good design practice to have a good understanding of the underlying manufacturing process. There are a number of general guidelines for the molding of precision optical components, as discussed throughout this chapter. It is important to remember that the precision glass molding process is really a compression molding process and very dissimilar to the injection molding process for optical plastics, though it uses some similar principles. Also, the variations in the process from supplier to supplier can easily impact what one supplier may prefer to another. It is always in the best interest of the designer to discuss its design with the manufacturer as early as possible in the design phase.

There are a number of standard lens shapes, all of which can be manufactured by precision glass molding. A precision glass molded lens can be biconvex, plano-convex, plano-concave, or meniscus. Biconvex and plano-convex lenses are the most popular. Utilizing a plano surface in design will simplify the manufacturing process, requiring less customized tooling and resulting in a lower-cost lens.

Lenses can be simplistically categorized into three sizes: small, standard, and large. Small lenses are those that have volumes less than the smallest readily available ball or gob preforms. If we consider the 0.75 mm diameter as the smallest readily available preform size, this is a volume of 0.22 mm^3. This would relate to a lens with an outside diameter of approximately 0.8 mm and a center thickness of 0.5 mm. Very small mold cavities can be difficult to manufacture. Small, steep surfaces may not be manufacturable. Many of these small lenses are made from larger prefroms and then postprocessed to get to the final form.

Standard size lenses are the most cost-effective, and correspond to readily available preform sizes. Assuming 1 to 8 mm ball preforms as readily available, this correlates to a typical lens with an outside diameter of approximately 8.0 mm and a center thickness of 5.5 mm. Outside of steep surfaces, molds are relatively easy to manufacture, thermal compensations are predictable, and processing times are reasonable.

Lenses larger than the standard size inevitably require custom preforms. Larger molds are more time-consuming to manufacture, and it becomes more difficult to predict and compensate for thermal compensations. The larger the lens, the longer the processing time will be. Costs will increase with size. The limitation on the size of lenses is driven by the molding equipment; maximum size is usually on the order of 100 mm diameter.

Molded lenses come in a variety of shapes and sizes; Figure 5.17 shows a number of different lenses. Table 5.5 shows the typical tolerances for a precision glass molded lens. Tolerances are dependent on many factors, though, and this chart should be used only as general guidance. Size and shape can have a significant impact on what is actually achievable.

5.7.1 Refractive Index Shift

As previously discussed, the molding process will have an impact on the refractive index and the Abbe number of the glass. The cooling rate of the molding process will be maximized by a molder in order to increase throughput of the process and reduce cost. These fast cooling rates are significantly faster than the fine anneal rates of the glass manufacturer. Hence, the slope on the T_g curve in Figure 5.5 will be shifted from the original glass melt. The result is a change in refractive index and dispersion from the published data from the glass suppler. The reduction is small: n_d

FIGURE 5.17
Variety of precision glass molded aspheric lenses.

TABLE 5.5

Typical Tolerances for a Precision Glass Molded Lens

Parameter	Commercial	Precision	Mfg. Limits
Center thickness, mm	±0.050	±0.025	±0.010
Diameter, mm	±0.025	±0.015	±0.005
Decentration, mm	±0.020	±0.010	±0.003
Wedge, arcmin	±10	±3	±1
Power/irregularity, fringes	5/2	3/1	0.5/0.25
Surface roughness, nm	20	10	5
Surface quality, scratch/dig	60-40	40-20	10-5

will drop by 0.002 to 0.005, and v_d drops by 0.1 to 0.3, depending on material and processing conditions. The refractive indices provided in Table 5.1 are the indices prior to molding. Individual molders will provide an as-molded index of refractive values based on their process; variation from manufacturer to manufacturer is very slight, but should always be confirmed. The as-molded values are typically determined by first developing a molding process for the specific glass and determining the annealing rate that will be used in production. This annealing rate is optimized in order to minimize cycle time while still producing a quality part. An equivalent molding process with annealing cycle is then conducted on a sample piece. A prism is fabricated from the sample, and the index of refraction is measured on a refractometer. The values measured are provided as the as-molded values.

A postannealing process could be used to further adjust the refractive index, but is typically unnecessary due the excellent repeatability of the molding process and risk of impacting the surface quality of the final product. The additional costs associated with postannealing also make it an undesirable option.

5.7.2 Diffractives

Diffractive features in molded glass exhibit many of the same issues discussed in Chapter 4 for injection molded diffractive surfaces. Some additional issues arise in glass molding, specifically related to the tooling. Grinding diffractive features in carbide or ceramic is very difficult due to the required shape of the grinding wheel. Grinding wheel designs have been developed to create these features, but the difficulty of manufacturing the tooling and the inherent diffractive efficiency losses in the visible range have limited the use of diffractives in precision glass molding. The exception to this is the molding of diffractive features in chalcogenide glasses. The advantages of diffractives in chalcogenide glass are addressed in Chapter 6.

5.8 AR Coating

There are no process-specific issues associated with coating precision glass molded lenses. All of the normal techniques and materials that apply to coating traditional ground and polished lenses also apply to molded lenses.

5.9 Summary

Precision glass molding is not a new process, though one that has been largely ignored throughout its development. Advancements in glass manufacture, preforms, molding equipment, and mold manufacture have moved precision glass molding to the mainstream. There are now over one hundred types of moldable glasses available, many meeting even the strictest of today's environmental regulations. This variety provides a significant number of options for optical designers. Moldable oxide and chalcogenide glasses are available for applications from ultraviolet all the way up to long-wave infrared wavelengths. Lenses as small as a millimeter and up to 100 mm in diameter can be molded. These lenses can include integral features for mounting, incorporate diffractive features, or even be insert molded into housings. Postprocessing can be used to generate unique features and geometries. Precision glass molding has become a cost-effective, high-quality, viable solution for even the most demanding of applications.

References

1. Webb, J. H., 1946, U.S. patent 2,410,616.
2. Angle, M. A, Blair, G. E., Maier, C. C., 1974, U.S. patent 3,833,347.
3. Parsons, W. F., Blair, G. E., Maier, C. C., 1975, U.S. patent 3,900,328.
4. Blair, G. E., Shafer, J. H., Meyers, J. J., Smith, F. T. J., 1979, U.S. patent 4,139,677.
5. Blair, G. E., 1979, U.S. patent 4,168,961.
6. Olszewski, A. R., Tick, P. A., Sanford, L. M., 1982, U.S. patent 4,362,819.
7. Marechal, J.-P., Maschmeyer, R. O., 1984, U.S. patent 4,481,023.
8. Hoya Optics, Inc., *Hoya optics product brief: Molded optical elements*, MOE-001, 6/90, Fremont, CA.
9. Toshiba Machine Company Ltd., Glass optical element made by high precision mold press forming. http://www.toshiba-machine.co.jp/
10. Fukuzaki, F., Ishii, Y., Fukuzawa, M., 1981, U.S. patent 4,249,927.
11. Yamane, M., Asahara, Y., 2000, *Glasses for photonics*, New York: Cambridge University Press.
12. Doremus, R. H., 1994, *Glass science*, New York: Wiley.
13. Schott North American, Inc., November 2007, *Technical information, advanced optics, TIE-40: Optical glass for precision molding*.
14. Hoya Corporation, USA Optics Division, 2007, Optical glass.
15. World Bank Group, 1998, *Pollution prevention and abatement handbook—Glass manufacturing*. http://smap.ew.eee.europa.eu/test1/fol083237/poll_abatement _handbook.pdf/
16. Reichel, S., Biertumpfel, R., 2010, Precision molded lens arrays made of glass, *Optik & Photonik*, June, No. 2.
17. Amo, Y., Negishi, M., Takano, J., 2000, Development of large aperture aspherical lens with glass molding, *Advanced Optical Manufacturing and Testing Technology 2000, SPIE*, 4231.
18. Katsuki, M., 2006, Transferability of glass molding, *SPIE*, 6149.
19. Cha, D., Hwang, Y., Kim, J., Kim, H., 2009, Transcription characteristics of mold surface topography in the molding of aspherical glass lenses, *Journal of the Optical Society of Korea*, 13 (2).
20. Maehara, H., Murakoshi, H., Quartz glass molding by precision glass molding method, Toshiba Machine Company Ltd. http://www.toshiba-machine.co.jp/

21. Hasegawa, M., et al., 2002, Optical characteristics of an infrared translucent close-packed ZnS sintered body, *SEI Technical Review*, 54 (June):71–79.

22. Minakata, T., Taniguchi, Y., Shiranaga, H., Nishihara, T., 2004, Development of far infrared vehicle sensor system, *SEI Technical Review*, 58 (June):23–27.

23. Uneo, T., et al., 2009, Development of ZnS lenses for FIR cameras, *SEI Technical Review*, October, 69:48–53.

24. Autery, W. D., Tyber, G. S., Christian, D. B., Barnard, M. M., 2007, U.S. patent 7,159,420 B2.

25. Autery, W. D., Tyber, G. S., Christian, D. B., Buehler, A. L., Syllaios, A. J., 2007, U.S. patent 7,171,827 B2.

26. Jeon, B. H., Hong, S. K., Pyo, C. R., 2002, Finite element analysis for shape prediction on micro lens forming, *Transactions of Material Processing*, 11(7).

27. CDGM Glass Co. Ltd., 2008, http://www.cdgmgd.com/en/asp/China

28. Hoya Corporation, Master optical glass datasheet. http://www.hoya-optical-world.com/english/datadownload/index.html.Japan

29. Ohara Corporation, Fine gobs for aspherical lenses. http://www.oharacorp.com/fg.html.Japan

30. Schott North American, Inc., 2009, Advanced optics, optical materials for precision molding.

31. Sumita Optical Glass, Inc., 2002, Optical glass data book, glass data version 3.03. http://www.sumita-opt.co.jp/data/Japan

32. LightPath Technologies, 2010, C0550 datasheet. http://www.lightpath.com/products/Orlando

33. LightPath Technologies, 2010, ECO-550 datasheet. http://www.lightpath.com/products/Orlando

34. Ohara Corporation, 2008, Low T_g optical glass for mold lenses. http://www.oharacorp.com/lowtg.html#fge

6

Molded Infrared Optics

R. Hamilton Shepard

CONTENTS

6.1 Introduction

Over the years production of infrared sensors and laser systems has increased dramatically and made infrared technology ubiquitous in a host of military, industrial, commercial, and medical applications. As new markets are discovered, demand continues to grow for affordable, high-quality optical systems operating in the infrared. Over the past decade molding technology has been used to successfully produce high-quality infrared lenses with surface complexities ranging from simple conics to complex aspheric diffractives. The topic of molded optics for infrared applications is becoming increasingly relevant as the affordability and availability of infrared detectors and light sources increase with every passing year.

In this chapter we provide an overview of the manufacturing process for molded infrared optics and a discussion of design considerations for their use. Applications for molded optics are considered for commonly used infrared wavebands: the short-wave infrared (SWIR, 0.9–1.7 μm), the mid-wave infrared (MWIR, 3–5 μm), and the long-wave infrared (LWIR, 8–12 μm).

Many of the moldable optical glasses discussed in Chapter 5, and to a lesser extent the moldable plastics of Chapter 4, are also viable candidate materials for the SWIR waveband. A brief section on design considerations for molded SWIR lenses is presented to highlight some of the unique challenges and opportunities related to selecting optimal lens materials in this waveband; however, the core discussion of their manufacturing process and practical use is left to the earlier chapters.

Within this chapter, an emphasis is placed on using molded chalcogenide lenses for MWIR and LWIR applications. The process for manufacturing molded chalcogenide lenses is generally consistent with that of molding glasses, as discussed in Chapter 5. In this chapter, manufacturing considerations specific for molded chalcogenides are addressed along with material properties, tolerances, surface quality and diffractive features, relative cost and affordability, and antireflection coatings.

The chapter continues with an overview of using molded chalcogenide optics in the MWIR and LWIR. The chromatic properties of molded chalcogenide lenses and their relationship to other infrared materials are presented along with important considerations for using diffractive surfaces in infrared systems. An example of a laser collimating lens is used to illustrate potential uses for molded chalcogenide optics in the MWIR. Molded lenses are of particular importance in the LWIR, where the availability of low-cost uncooled detectors has provided the impetus to develop affordable lenses that are producible in high volumes. Examples of common LWIR lens designs are provided, and comparisons are drawn between molded and traditional lenses to highlight the relative image quality, manufacturability, and thermal performance of these systems.

6.2 Using Molded Optics in the Short-Wave Infrared

The short-wave infrared (SWIR) is a waveband that spans the gap between the visible and the thermal infrared. A SWIR imager observes the confluence of the familiar scene phenomenology of objects observed by reflected light as well as thermal signatures from very hot objects. For some time the SWIR waveband has been used for infrared astronomy, remote sensing, and military applications, and in recent years an increased availability of low-noise, uncooled SWIR detectors has sparked new interest for applications such as machine vision, agricultural inspection, and next-generation night vision technology. The SWIR band is also home to a number of laser systems used for telecommunications, medical, and military applications. As applications and production volumes increase, the affordability of low-cost moldable lenses will become increasingly relevant.

In practice, the extent of the SWIR waveband is often defined by the detector technology with which it is used. For example, this might be 1 to 2 microns for a cooled InSb detector, or 0.9 to 1.7 microns for an uncooled InGaAs detector. In the present discussion we define the SWIR in accordance with the latter, since uncooled SWIR detectors have a greater potential to be lower cost, and are therefore more germane to a discussion of molded optics. With a few considerations, a lens designer has a wide range of material options available in this regime, including the use of molded optics.

Material choices abound in the SWIR. The majority of optical glasses traditionally used for visible applications are also transparent out to around 2 microns. This includes many of the moldable glass materials discussed in Chapter 5. Additionally, infrared materials such as zinc sulfide and zinc selenide offer a high refractive index and low dispersion, which is very useful for aberration correction.

Infrared chalcogenides provide another option for high-index materials in the SWIR. Many chalcogenides, including some moldable varieties, are reported to be transparent down to below 1 micron. Additionally, the molding process can produce surfaces with roughness comparable to those of traditional polishing, so that scatter at the shorter SWIR wavelengths is not an issue. However, for imaging applications, the lens designer should be cautious when using an amorphous material in close proximity to its absorption edge. It is advisable to evaluate the risk of potential melt-to-melt variations in the refractive index or other optical properties. If system requirements allow, it might be worthwhile to investigate trading a reduced waveband for a faster F/number.[1] Moving the operational waveband away from the absorption edge of chalcogenides and compensating the transmission loss with a faster system can allow chalcogenides to be used with relatively low risk.

For laser applications, a designer may consider using molded plastic lenses for collimators and beam expanders. Many of the molded polymer materials discussed in Chapter 4 are transparent at certain laser wavelengths within the SWIR. It might also be tempting to use moldable plastic optics for SWIR imaging applications where weight is a chief concern. However, the presence of multiple strong absorption bands within the SWIR waveband will likely preclude their use in broadband applications. Figure 4.3 can be consulted for more information on polymer absorption in the SWIR.

When selecting moldable materials for lenses operating in the SWIR, the designer should be aware of three challenges related to correcting chromatic aberration: (1) the relative width of the waveband with respect to its central wavelength is broader than most other visible and infrared wavebands, (2) the roll-off in diffraction efficiency vs. wavelength might preclude using diffractive surfaces for color correction in the SWIR, and (3) many of the available optical glasses exhibit lower dispersion in the SWIR than in the visible, which makes it difficult to find strong flints for achromatization. Figure 6.1 shows the relative locations of selected moldable glasses on the n-*V* diagram

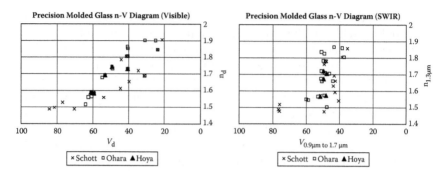

FIGURE 6.1

Refractive index vs. dispersion for selected precision molded glasses. Left: Optical properties in the visible. Right: Optical properties of the same glasses in the 0.9 to 1.7 μm SWIR waveband.

in the visible and in the SWIR. In the SWIR these moldable glasses are perfectly viable. However, with little variety in material dispersion, achieving achromatization may cause lens elements to become very strong. With stronger elements come more steeply curved surface radii, which can increase the aberration content of a lens. Fortunately, the option of using molded aspheric surfaces provides a great benefit in this area.

The design and implementation of a molded SWIR lens is much like its visible counterpart. Many of the moldable optical glasses discussed in Chapter 5 can be used effectively in SWIR applications. Color correction can be a challenge in the SWIR, but if necessary, certain infrared materials may be used. Caution should be exercised when using moldable chalcogenides in the lower SWIR waveband. Moldable plastic optics are viable for laser applications, but should be avoided for broadband SWIR imagers.

6.3 Manufacturing Considerations for Molded Infrared Optics

Precision molding of chalcogenide glass requires special attention. Chalcogenide glasses predominantly contain one or more of the chalcogenide elements: sulfur, selenium, or tellurium. Chalcogenide elements are defined in the periodic table as subgroup VI-A, shown in Table 6.1. Selenium is the

TABLE 6.1

The Chalcogens

Order No.	Element	Symbol
16	Sulfur	S
34	Selenium	Se
52	Tellurium	Te

TABLE 6.2

Material Compositions of a Selection of Chalcogenide Glasses for
Precision Molding

Grade	Manufacturer	T_g (°C)	Approximate Elemental Composition (%)				
			Ge	Sb	Se	As	Te
IG2	Vitron	368	~33	0	~55	~12	0
IG3	Vitron	275	~30	0	~32	~13	~25
IG4	Vitron	225	~10	0	~50	~40	0
IG5	Vitron	285	~28	~12	~60	0	0
IG6	Vitron	185	0	0	~60	~40	0
GASIR 1	Umicore	292	~22	0	~68	~20	0
GASIR 2	Umicore	263	~20	~15	~65	0	0
AMTIR-1	AMI	368	~33	0	~55	~12	0
AMTIR-3	AMI	278	~28	~12	~60	0	0
AMTIR-5	AMI	143	0	0	~64	~36	0

primary chalcogenide component in almost all commercially available chalcogenide materials for precision glass molding. The approximate material compositions of several of these materials are shown in Table 6.2, along with their glass transition temperatures.

The most well-known suppliers of chalcogenide glasses for precision molding are Amorphous Materials,[2] Umicore,[3] and Vitron.[4] Each has its own group of material grades with several overlapping compositions; for example, AMI's AMTIR-1 has the same composition as VITRON's IG-2. AMTIR-1/IG2 and AMTIR-3/IG5 were originally developed by Texas Instruments as TI-20 and TI-1173, respectively. There are a significant number of other chalcogenide glasses available in the commercial market for other applications; the ones listed simply meet the definition of moldable glass, as discussed in Chapter 5.

One of the largest differences between molding chalcogenide and the oxide-based glasses of Chapter 5, besides composition, is the cost of the material itself. For many of the popular oxide glasses the contribution of the raw material is almost negligible in the overall cost of a lens. Conversely, for a molded chalcogenide lens the raw material can represent a significant portion of the overall cost. Therefore, efficient use of material becomes imperative in the manufacture of chalcogenide glass lenses. When compared to germanium, chalcogenide glasses can present a significant cost advantage. While many chalcogenide glasses contain germanium, it is usually less than a third of the overall composition. Germanium is the most expensive component in moldable chalcogenide glasses and can be a driving factor in the material cost. Table 6.3 shows the 2008 pricing of some of the most common raw materials used in moldable chalcogenide glasses.

A simplified cost comparison based on material percentages is shown in Table 6.4. The raw material cost of a chalcogenide glass such as IG6, which

TABLE 6.3

2008 Average Raw Material Costs
of Selected Elements Used in the
Manufacture of Chalcogenide Glass

Element	Symbol	Cost
Germanium[a]	Ge	$1,600.00/kg
Selenium	Se	$72.75/kg
Tellurium	Te	$215.00/kg
Antimony	Sb	$6.26/kg
Arsenic	As	$2.16/kg

Source: U.S. Geological Survey, Commodity
Statistics and Information, http://
minerals.usgs.gov/minerals/
pubs/commodity/, 2010.[22]

[a] Germanium cost is year-end for zone
refined.

TABLE 6.4

Estimated Raw Material Costs of Selection of Chalcogenide Glasses

Material	Material Composition				Total Raw Material Cost of 1 kg	Relative Cost to 1 kg of Ge
	Ge	Sb	Se	As		
Ge	100%	0%	0%	0%	$1,600.00	100%
	$,1600.00	$0.00	$0.00	$0.00		
IG2	33.00%	0.00%	55.00%	12.00%	$576.74	36.0%
	$528.00	$0.00	$40.01	$8.73		
IG5	28.00%	12.00%	60.00%	0.00%	$492.40	30.8%
	$448.00	$0.75	$43.65	$0.00		
IG6	0.00%	0.00%	60.00%	40.00%	$72.75	4.5%
	$0.00	$0.00	$43.65	$29.10		

does not contain germanium, is one-twentieth that of pure germanium. Obviously, this is just a raw material cost; it does not include the processing costs, environmental costs, capital equipment, etc. These prices show what might be achievable in chalcogenide glass manufacture; current pricing does not reflect this. This is because the manufacture of chalcogenide glass is still a rather immature technology with limited volumes. As the volume of use increases and industry adoption takes place, one would expect the economies of scale in production, industry competition, and improvements in efficiencies to drive prices to approach this level. This is why we see chalcogenide glasses used predominantly in high-volume designs for automotive use.[5,6]

The reduction of design dependency on the use of germanium greatly reduces the cost sensitivity of a system. Germanium has seen significant pricing fluctuations in recent years. The cost of germanium doubled from

2006 to 2008 prior to returning to 2006 values in 2010.[7] Thus, a designer look-ing to insulate their design from volatile price fluctuations might choose a chalcogenide glass over germanium.

Compared to oxide glasses, cost is not the only difference; chalcogenide glasses possess significantly different mechanical, thermal, and optical prop-erties. Chalcogenide glasses have lower mechanical strength and thermal stability than the oxide glasses and much higher coefficients of thermal expan-sion. The mechanical and thermal properties of several chalcogenide glasses are shown in Table 6.5. The primary noticeable difference between moldable chalcogenide glasses and the moldable oxide glasses in Table 5.2 is the differ-ence in glass transition temperature, T_g. The T_gs of most of the commercially available chalcogenides are shown in Table 6.5. The average T_g of the visible glasses is approximately 500°C, whereas the moldable chalcogenide glasses have an average of 240°C, a significant difference of over 250°C.

The molding of chalcogenide lenses follows the same principles and pro-cedures as discussed in Chapter 5. Changes in the molding process must, however, take into account the differences in thermal properties between the oxides and the chalcogenides; this includes changes to the molds and the molding equipment. Chalcogenide glasses are typically melt cast in rocker furnaces in quartz tubes; therefore, the availability of the bulk material is typically in larger boule form. These boules are usually on the order of 100 to 200 mm in diameter, with lengths ranging from 50 to 200 mm. Preforms are manufactured from the boules and subsequently molded into lenses. Typical manufacturing tolerances for chalcogenide lenses are provided in Table 6.6.[8]

It can be advantageous to mold diffractive features onto a chalcogenide lens. Figure 6.2 shows a diffractive surface molded onto a chalcogenide glass lens. The theory behind using diffractive surfaces and the process by which they are molded onto chalcogenide lenses is the same as discussed in ear-lier chapters. One distinct advantage of using molded diffractive surfaces on infrared optics is that the longer wavelength makes the elements less sensi-tive to tolerance errors in the diffractive profile and step height. The issues discussed in previous chapters on molded diffractives are the same, but the reduced sensitivity to fabrication errors has the potential to make diffrac-tive surfaces more efficient and viable for molded optics in the MWIR and LWIR wavebands.

As previously discussed, reducing the germanium content of a chalco-genide glass can provide a significant cost advantage. However, as the ger-manium content is reduced, so too is the glass transition temperature.[9] A lower glass transition temperature can be advantageous for the molding pro-cess, but it can complicate the processes by which the materials are coated. Lower temperatures are required for coating molded chalcogenide optics, and the softness of the material can make the process challenging. Processes such as low-temperature plasma deposition[10] and ion-assisted deposition[9] have proven effective in producing durable, high-efficiency antireflective coatings in both the MWIR and LWIR. Thorium-free antireflective coatings

TABLE 6.5

Selected Mechanical and Thermal Properties of Commercially Available Chalcogenide Glasses

	Density g/cm³	Thermal Expansion 10⁻⁶ K⁻¹ @ 20–100°C	Specific Heat J/gK	Thermal Conductivity W/mK	Knoop Hardness —	Transition Temp. T_g °C	Softening Point A_t °C
IG 2	4.41	12.1	0.33	0.24	141	368	445
IG 3	4.84	13.4	0.32	0.22	136	275	360
IG 4	4.47	20.4	0.37	0.18	112	225	310
IG 5	4.66	14.0	0.33	0.25	113	285	348
IG 6	4.63	20.7	0.36	0.24	104	185	236
AMTIR-1	4.40	12.0	0.30	0.25	170	368	405
AMTIR-2	4.66	22.4	0.36	0.22	110	167	188
AMTIR-3	4.67	14.0	0.28	0.22	150	278	295
AMTIR-4	4.49	27.0	0.36	0.22	84	103	131
AMTIR-5	4.51	23.7	0.28	0.24	87	143	170
AMTIR-6	3.20	21.6	0.46	0.17	109	187	210
GASIR 1	4.40	17.0	0.36	0.28	194[a]	292	N/A
GASIR 2	4.70	16.0	0.34	0.23	183[a]	264	N/A
GASIR 3	4.79	17.0	N/A	0.17	160[a]	210	N/A

[a] Converted from Vickers microhardness per ASTM E140-07.

TABLE 6.6

Typical Manufacturing Tolerances
of Molded Chalcogenide Lenses

Parameter	Typical Value
Center thickness (CT)	±0.025 mm
Diameter	±0.010 mm
Decentration	±0.005 mm
Wedge	1 arcmin
Power/irregularity	1/0.5–0.5/0.25
	Fringes @ 633 nm
Surface roughness	10 nm
Scratch/dig	60–40 to 40–20
Refractive index	±0.030 mm

FIGURE 6.2
A diffractive surface molded onto a chalcogenide lens.

for chalcogenide glasses have achieved average transmissions greater than 98% for the 8 to 12 μm wavelength band and pass the moderate abrasion, humidity, and adhesion requirements of MIL-C-48497A.[6] Coatings have also been developed to improve the durability of the lenses, specifically for exposed first-surface applications where diamond-like carbon (DLC) coatings are typically used.[11]

6.4 Design Considerations for Molded Infrared Lenses

Certain infrared lens layouts have stood the test of time by providing superior performance to other alternatives. Such examples frequently reoccur in modern designs and include the MWIR silicon germanium doublet and the LWIR germanium ZnS doublet. These lenses credit their performance to the high refractive index and wide variation in the dispersion of available infrared materials. In fact, the dispersion of germanium is so low in the LWIR that in some cases color correction can essentially be ignored, giving the designer the luxury of correcting monochromatic aberrations with an $n = 4$ material. By comparison, the optical properties of chalcogenide glasses might seem bland; their refractive index is not as high as germanium or silicon, nor is their dispersion extreme enough to deem them ideal candidates for color correction. At first glance, they do not necessarily lend themselves to novel lens designs that would outperform the traditional mainstays. However, a designer need not look far to find compelling reasons to consider them as alternatives to designs that rely on germanium optics.

Molded chalcogenide materials introduce the potential for significant cost savings, the added flexibility of having more infrared materials to work with, and the possibility of optomechanical benefits, such as molding lenses into metal structures[12] for edge protection or to provide mounting features. Moreover, superior thermal performance is possibly the most significant advantage of using chalcogenide lenses (molded or otherwise). Germanium is notoriously sensitive to thermal changes. Its thermoptic coefficient (dn/dt) is at least twice as large as any other common infrared material, and many times larger than some. This leads to a relatively large shift of the image plane with changing temperature that can complicate passively athermal-izing a fixed-focus lens, particularly when the focal length is long. By comparison, the thermoptic coefficient of typical chalcogenides is on the order of five times less than that of germanium, which can be used to alleviate the challenges imposed by thermal focus shift. Another undesirable thermal property of germanium is thermal darkening, which limits its useful temperature range to less than 80°C. By contrast, a chalcogenide lens can be useful at temperatures of 100°C or higher, depending on its glass transition temperature. It has been shown that in some situations germanium elements can be substituted with chalcogenide to enhance the thermal performance of an MWIR lens.[10] Furthermore, certain moldable chalcogenides such as AMTIR-5 have been developed with specific thermal properties, such as a near-zero refractive index change with temperature and a coefficient of thermal expansion to match that of an aluminum lens housing.

Chalcogenide lenses have been used for decades,[13,14] and their virtues and limitations are well established in the literature. High-quality infrared lenses such as those shown in Figure 6.3 can be constructed entirely from molded chalcogenide elements and are currently available from commercial suppliers. Alternatively, chalcogenide glasses can be used in conjunction with traditional infrared materials to provide improvements to cost and thermal performance.[8] The present discussion focuses on applications that maximize the advantages of using molded chalcogenides. After an overview of the chromatic properties of molded chalcogenides, an example of using

FIGURE 6.3
Commercially available LWIR lenses constructed from molded chalcogenide glasses.

a molded chalcogenide lens for an MWIR laser collimator is presented. A detailed discussion of LWIR lens designs follows. More emphasis is placed on LWIR lens designs because the abundance of lower-cost uncooled micro-bolometers increases the likelihood of high-volume applications, and the improved thermal performance of chalcogenide lenses makes them promising alternatives to traditional germanium-based designs.

6.4.1 Chromatic Properties of Molded Infrared Materials

In complex multielement lens systems subtle differences between the chromatic properties of available materials can often make the difference in whether or not a design is feasible. A lens designer working in the thermal infrared has relatively few options for good materials that provide high transmission as well as robust chemical and physical durability. Prior to the advent of infrared chalcogenide glasses, a designer working in the LWIR could count the commonly used lens materials on one hand: germanium with an exceptionally high index and low dispersion, zinc sulfide and zinc selenide with higher dispersion, making them good flints, and gallium arsenide. A few more material options exist for the MWIR: the aforementioned materials (albeit with different dispersion) plus silicon (the best MWIR crown), and a few lower-index/higher-dispersion materials, such as calcium fluoride, magnesium fluoride (polycrystalline), and sapphire. Infrared chalcogenides are now available from multiple suppliers, and slight variations in their optical properties have increased the number of unique infrared materials by two-to threefold.

Today's moldable infrared materials are amorphous chalcogenide glasses in which the refractive index and dispersion are determined by the particular blend of constituent materials. This is unlike most other infrared materials that derive their properties from a predetermined crystalline structure. Chalcogenide glasses can be tailored to some degree, and thus moldable varieties have been developed to closely match the optical properties of

TABLE 6.7

Optical Properties of Selected Chalcogenide Glasses

Trade Name	Composition	Refractive Index	Effective Abbe Number	dn/dt ($\times 10^{-6}$/K)
AMTIR-1	Ge-As-Se	2.5146 @ 4 μm	202 3–5 μm	86 @ 3.4 μm
		2.4981 @ 10 μm	109 8–12 μm	72 @ 10 μm
AMTIR-2	As-Se	2.7760 @ 4 μm	171 3–5 μm	31 @ 4 μm
		2.7613 @ 10 μm	149 8–12 μm	5 @ 10 μm
AMTIR-3	Ge-Sb-Se	2.6216 @ 4 μm	159 3–5 μm	98 @ 3 μm
		2.6027 @ 10 μm	110 8–12 μm	91 @ 10 μm
AMTIR-4	As-Se	2.6543 @ 4 μm	186 3–5 μm	−24 @ 3 μm
		2.6431 @ 10 μm	235 8–12 μm	−23 @ 10 μm
AMTIR-5	As-Se	2.7545 @ 4 μm	175 3–5 μm	<1 @ 4 μm
		2.7398 @ 10 μm	172 8–12 μm	<1 @ 10 μm
AMTIR-6	As-S	2.4107 @ 4 μm	155 3–5 μm	<1 @ 5 μm
		N/A @ 10 μm	N/A 8–12 μm	N/A @ 10 μm
IG2	Ge-As-Se	2.5129 @ 4 μm	201.7 3–5 μm	71 @ 3.4 μm
		2.4967 @ 10 μm	105.4 8–12 μm	60 @ 10.6 μm
IG3	Ge-As-Se-Te	2.8034 @ 4 μm	152.8 3–5 μm	130 @ 4 μm
		2.7870 @ 10 μm	163.9 8–12 μm	145 @ 10.6 μm
IG4	Ge-As-Se	2.6210 @ 4 μm	202.6 3–5 μm	30 @ 3.4 μm
		2.6084 @ 10 μm	174.8 8–12 μm	36 @ 10.6 μm
IG5	Ge-Sb-Se	2.6226 @ 4 μm	180.3 3–5 μm	76 @ 3.4 μm
		2.6038 @ 10 μm	102.1 8–12 μm	91 @ 10.6 μm
IG6	As-Se	2.7945 @ 4 μm	167.7 3–5 μm	35 @ 3.4 μm
		2.7775 @ 10 μm	161.6 8–12 μm	41 @ 10.6 μm
GASIR 1	Ge-As-Se	2.5100 @ 4 μm	196.1 3–5 μm	84 @ 1.06 μm
		2.4944 @ 10 μm	119.6 8–12 μm	55 @ 10.66 μm
GASIR 3	Ge-Sb-Se	2.6287 @ 4 μm	169.7 3–5 μm	53 @ 10.66 μm
		2.6105 @ 10 μm	115.0 8–12 μm	

legacy chalcogenides such as AMTIR-1/TI-20 and AMTIR-3/TI-1173. Physical durability and thermal properties are additional considerations impacting the design of chalcogenide glasses. Table 6.7 provides a list of selected optical properties for commercially available chalcogenide glasses. The table includes published information for some of the most commonly used infrared chalcogenides. It should be noted that some suppliers of molded lenses apply a proprietary trade name to their finished product. The industry is continually working to improve and develop new moldable chalcogenide glasses, and so it is recommended that material availability be confirmed with the supplier prior to beginning a lens design.

Designing a lens using a molded IR material is fundamentally the same as designing with any other material. The lens designer typically selects a combination of materials in which the dispersive properties of positive and negative elements counteract each other to achieve an achromatized solution. An

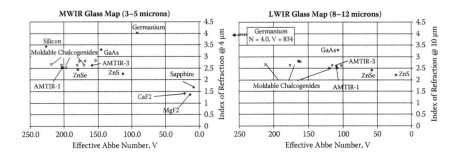

FIGURE 6.4
Mid-wave infrared and long-wave infrared glass maps. Moldable chalcogenide materials are plotted with open circles.

n-V diagram is a common tool used to observe the chromatic properties of a material. In the infrared the material's refractive index, n, is plotted against its effective infrared Abbe number, V, defined as follows:

$$V = \frac{n_{\lambda 2} - 1}{n_{\lambda 1} - n_{\lambda 3}} \tag{6.1}$$

where $n_{\lambda 1,2,3}$ is the refractive index at the respective wavelengths $\lambda 1$, $\lambda 2$, and $\lambda 3$, arranged in ascending order. To first order, an achromatic doublet consists of a positive lens with a low dispersion (high V) and a weaker negative lens with a higher dispersion (low V). The power required for the individual elements is inversely related to the difference in their effective Abbe numbers, and so choosing materials with a large horizontal spacing on the n-V diagram is most desirable. n-V diagrams for materials in the MWIR and LWIR are shown in Figure 6.4. Some of the most commonly used infrared materials are shown. Many of the materials not shown have undesirable characteristics, such as poor chemical or physical durability, and are therefore used less frequently. The addition of chalcogenides to the infrared design space provides a valuable new set of material options. Their refractive index is high enough to function as efficiently as ZnS and ZnSe while providing a useful alternative in terms of their dispersion.

Transmission can be an important consideration when selecting a moldable chalcogenide glass. The majority of these materials have very low bulk absorption in the 3 to 5 μm MWIR region, and so the transmission of a lens is essentially only limited by its antireflective coating. However, a long-wavelength absorption edge present in many chalcogenide glasses can limit the transmission of lenses operating in the LWIR. Figure 6.5 shows the theoretical transmission of uncoated 10 mm thick chalcogenide windows. In the MWIR transmission loss results primarily from Fresnel reflection; however, at longer wavelength transmission is limited by bulk absorption.

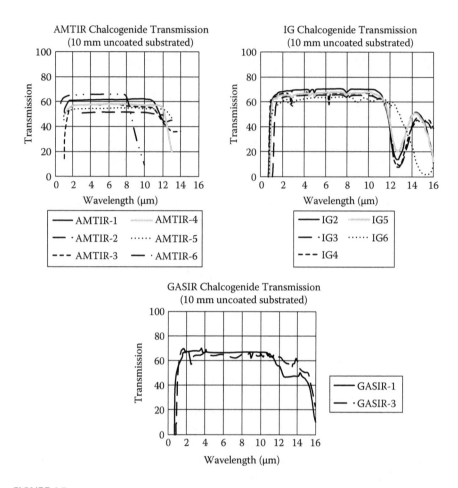

FIGURE 6.5
Transmission of selected commercially available chalcogenides.

6.4.2 Considerations When Using Diffractive Surfaces

When designing molded infrared lenses the higher dispersion of chalco-genides might warrant diffractive optical elements to correct chromatic aberrations. As discussed earlier, the molding process can produce diffractive kinoform surfaces with accurate step heights and sharp transition zones. Diffractive surfaces are common in infrared systems and can improve image quality significantly. However, it is important for the designer to be aware of the risks and trade-offs associated with their use. Chapter 3 discussed stray light risks introduced by diffractive surfaces such as scattering and ghost images. Infrared systems can be particularly sensitive to these ghost images since diffractive rings can act as thermally emissive sources within the field of view. Although the transmission of the ghost path will be relatively low compared to the overall system transmission, the temperature difference between the lens and the scene may cause rings to appear in the image plane.

Another potential concern is the transmission and modulation transfer function (MTF) loss related to diffraction efficiency, which was discussed in Chapter 1. The designer must consider these factors since they often are not observed in the lens design software. Note that a diffractive surface operating in the 8 to 14 μm waveband has almost twice the efficiency loss of an 8 to 12 μm diffractive. Additionally, it might be tempting to use two or more diffractive surfaces to achromatize an optical design, but doing so causes the efficiency losses of each surface to be multiplied.

Applying diffractive surfaces to molded infrared lenses has been shown to be an effective means of correcting chromatic aberrations.[15] However, the performance increase comes at a price, and so diffractives should be used sparingly and with careful consideration.

6.4.3 Mid-Wave Infrared Applications for Molded Infrared Optics

Chalcogenide lenses have been used in mid-wave infrared (MWIR) applications for decades. The recent boom in moldable infrared technology has increased the abundance and variety of chalcogenide glasses available to produce lenses. For many MWIR imagers an actively cooled infrared detector package is often the largest cost driver for the system. Therefore, the advantage in cost reduction realized by using a molded lens might only be a small fraction of the overall system cost. Still, there are other benefits. New moldable chalcogenides not only increase the variety of materials available for MWIR color correction, but also provide an additional degree of freedom for athermalizing a lens.

Not only are infrared imagers seeing increased usage, but so too are infrared laser systems. A wide range of technologies have been developed based on infrared laser diodes and gas lasers, and most recently new quantum cascade lasers are spurring MWIR applications such as infrared spectroscopy, environmental monitoring and gas detection, explosives detection, and infrared countermeasures. To keep pace with the times, molding technology is now being used to create low-cost, single-element chalcogenide collimating lenses that can be used with MWIR laser sources.

When designing a single-element laser collimator aspheric surfaces are used to eliminate spherical aberration and produce diffraction-limited image quality over a small field of view. Although a singlet cannot be achromatized without the aid of a diffractive, optimizing the lens over a broader waveband will help control spherochromatism and allow the collimator to be used at multiple laser wavelengths (with a focus shift). Figure 6.6 shows an example of an MWIR laser collimator that could be produced with a molded infrared material. It consists of a single aspheric surface with parallel flats on either side for mechanical mounting. Image quality is better than

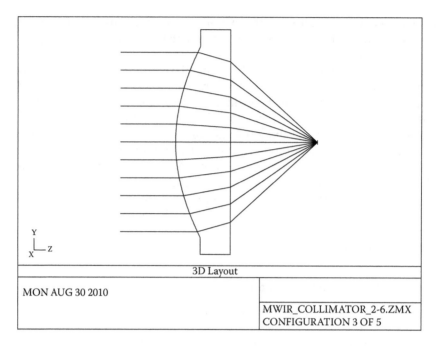

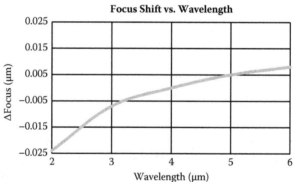

FIGURE 6.6
A moldable MWIR laser collimator with a 2 mm focal length and a numerical aperture of 0.8.

one-tenth wave peak to valley when coupled to an infrared laser in the 2- to 6-micron waveband.

6.4.4 Long-Wave Infrared Applications for Molded Infrared Optics

Germanium has long been the workhorse of LWIR lens design. For a system to harness the potential cost and thermal benefits of using molded chalcogenide, the designer must make the case that a molded design can rival the performance of a germanium-based lens without added size or complexity.

Astute selection of aspheric and diffractive surfaces allows molded infrared lenses to perform well for many LWIR applications. With a few considerations, designing single-FOV lenses for uncooled LWIR detectors can be a relatively straightforward process.

System requirements play the largest role in determining the final design form. Today's uncooled microbolometers typically require lenses operating in the range of F/0.8 to F/1.4 to achieve desirable thermal sensitivity (i.e., noise equivalent temperature difference on the order of 40 to 75 mK). Many of these sensors are advertised with a spectral response of 8 to 12 μm,[16] yet others specify a response in the range of 7.5 to 13.5 μm[17,18] or 8 to 14 μm.[19] In practice, the difference in image quality of a lens operating from 8 to 12 μm vs. 8 to 14 μm is often small. However, the lens designer should be cognizant of the operational waveband, as it can affect optimization for chromatic aberrations, the design and performance of diffractive surfaces, and material selection, since some chalcogenide glasses transmit poorly above 12 μm.

The fast F/numbers needed for uncooled detectors produce a diffraction-limited image resolution much higher than the limiting resolution of today's uncooled detectors. Uncooled LWIR detectors are currently available with pixel sizes ranging from 17 to 50 microns, producing Nyquist frequencies in the range of 10 to 30 cycles/mm. However, at a wavelength of 10 microns an F/1 lens has a diffraction-limited spatial frequency cutoff of 100 cycles/mm. This effectively means that an LWIR lens can tolerate higher levels of aberration than would be traditionally allowed in diffraction-limited systems, and that MTF at the lower spatial frequencies will provide a more relevant merit function than wavefront or ray aberration.

In terms of focal length, an LWIR design much shorter than 10 mm can be a challenge due to the back focal distance requirement of the camera package. Uncooled cameras typically require a back focal distance long enough to accommodate a shutter mechanism used for nonuniformity correction. In short focal length lenses this requirement can drive the back focal distance to be greater than the effective focal length of the lens. The upper bound for focal length in an LWIR lens is more of a practical consideration. Leading elements in long focal length designs can become very large due to the F/number requirement. Diffractive surfaces can also become an issue because the number of zones required to correct axial color typically scale with focal length.

Another consideration that drives the design form is the detector array size or field or view. Many commercially available uncooled detectors range in size from 160 × 120 pixels to 1,024 × 768 pixels. For a wide range of focal lengths, good image quality can be achieved in an LWIR lens using only two elements; however, when the detector array is relatively large with respect to the lens's focal length, some of the shorter focal length LWIR lenses are driven to three-element solutions.

Of the most common layouts employed for LWIR lenses, two of the classical formats will be presented to illustrate important design considerations. A retrofocus design is often used to provide a long back focal distance for short focal length lenses. When a longer focal length is required, the size of lens elements becomes considerable, and so a Petzval layout is often preferred. To assess the relative advantages and disadvantages of molded infrared lenses, comparisons are made to traditionally fabricated germanium lenses.

6.4.4.1 The Molded Infrared Retrofocus Lens

When the back focal distance requirement approaches or exceeds the system focal length, the retrofocus (a.k.a reverse telephoto) serves as an effective layout. Molded infrared retrofocus lenses have been demonstrated to rival the image quality of traditionally fabricated germanium lenses while showing improvements to the thermal performance.[20]

The underlying principle of a retrofocus arrangement is that when a negative lens is placed at the front focal point of a positive lens, the principal plane locations are shifted without changing the overall focal length. This creates a back focal distance that is longer than the effective focal length. The aperture stop is often placed at the front element. Placement of the stop near the front focal point of the second lens drives the system toward telecentricity in image space, minimizing the off-axis chief ray angle at the image plane. This provides the benefit of improving the uniformity of image irradiance by reducing the so-called Cos^4 roll-off that often plagues wide field of view systems.

Asymmetry of the layout complicates aberration correction. In theory, correction of chromatic aberrations requires that the positive and negative components be individually achromatized, and so in visible systems these lenses often become multielement groups. However, in the LWIR the relatively low dispersion of molded chalcogenide lenses and the low Nyquist frequency of the uncooled detectors allow chromatic aberration to be corrected with a single diffractive surface. Lateral color can be an issue in lenses with a wide field of view. As an "odd" aberration (having odd-powered pupil dependence), lateral color will be controlled most efficiently by placing the diffractive surface where the chief ray is highest: at the second lens. Correcting coma is typically difficult in asymmetric layouts, but the burden is relieved with liberal application of aspheres and the relatively low threshold for aberration correction necessary to produce good MTF below the Nyquist frequency.

Field curvature and distortion can present a challenge in a retrofocus layout. Aspheric surfaces have no impact on field curvature, and so controlling the Petzval sum is difficult when using only moldable chalcogenide lenses since the available materials all have a similar refractive index. A slight improvement is possible by using a higher-index chalcogenide ($n = 2.6$ to 2.7) for the stronger positive lens and a lower-index chalcogenide ($n \approx 2.5$) for the weaker negative lens. However, this modest attempt to correct field curvature is generally insufficient to allow lens powers and astigmatism (used to balance field curvature) to optimize unconstrained. Distortion can become significant in a retrofocus lens due to its asymmetric layout. One method to

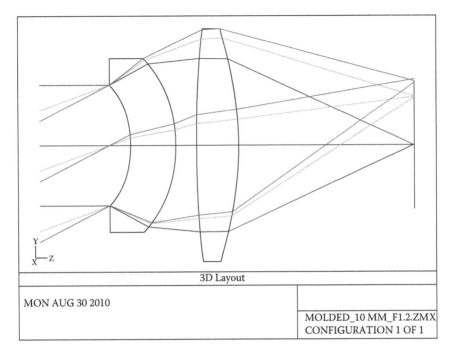

FIGURE 6.7
An F/1.2 molded chalcogenide retrofocus lens with a 10 mm focal length.

reduce distortion is to place an aspheric surface close to the image. However, this has its limits, and in wide field of view systems, distortion can become significant.

As an example of a moldable chalcogenide retrofocus lens Figure 6.7 shows a 10 mm focal length F/1.2 system operating in the wavelength range of 8 to 12 μm. The back focal distance was constrained to be greater than 12.5 mm to accommodate the protective window and shutter on the camera package, which drove the lens to the retrofocus layout. An image format of 320 × 240 with 25 μm pixels provides a full horizontal FOV of 43.6°. This set of requirements was selected as an example of a system that might be used for vehicle night vision or helmet-mounted applications where horizontal FOVs in the vicinity of 40° are typical.

The first element uses a Ge-As-Se chalcogenide glass, GASIR 1, with a refractive index of 2.4944 at 10 μm. The second lens provides the positive focusing power, and so it uses a higher refractive index to help reduce the Petzval sum. Its material is GASIR 3 (Ge-Sb-Se) with a refractive index at 10 μm of $n = 2.6105$. The stop is located at the first element, and its physical aperture is provided by the lens seat. In this design example it was necessary to make every surface aspheric to achieve adequate aberration correction. A kinoform diffractive surface with seven zones is used on the rearmost surface to provide maximum control over lateral color.

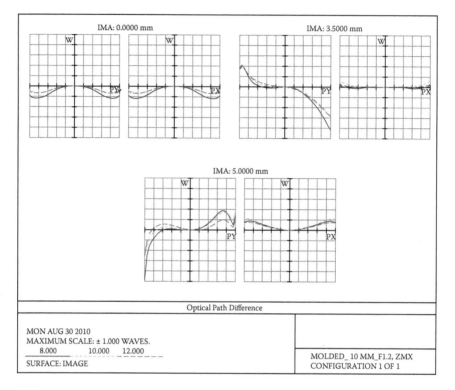

FIGURE 6.7 (continued)

A closer look at the transverse wavefront aberration plot at the upper right of Figure 6.7 indicates that coma is the dominant aberration in the system, which is expected from the asymmetric layout. Distortion was constrained to less than 9% in the optimization process, and further reduction was found to have diminishing return with respect to image quality. As mentioned previously, the relatively low sampling resolution of currently available uncooled detectors allows the system to tolerate wavefront error approaching one wave peak to valley, while still providing nearly diffraction-limited performance below the Nyquist frequency.

To assess the relative benefits of molded infrared lenses, a traditionally fabricated lens is presented in Figure 6.8 for comparison. The lens consists of two germanium elements. In this example similar performance to the molded lens is achieved using only one aspheric surface, but the MTF can easily be brought to the diffraction limit by simply increasing the number of aspheres. The low dispersion of germanium allows for adequate color correction without the use of diffractive elements.

The most significant difference between the molded and traditional lens relates to their thermal performance. A common operational temperature range for commercially available lenses is –40 to +80°C. Assuming the lenses

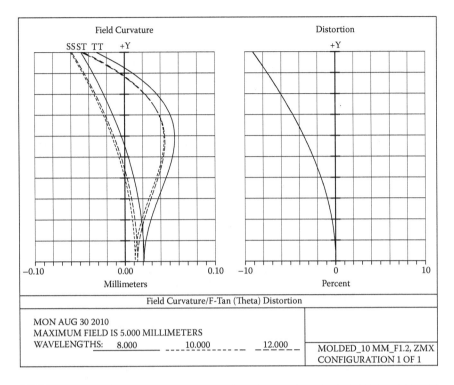

Field Curvature/F-Tan (Theta) Distortion

MON AUG 30 2010
MAXIMUM FIELD IS 5.000 MILLIMETERS
WAVELENGTHS: 8.000 ——— 10.000 —————— 12.000 ———

MOLDED_10 MM_F1.2, ZMX
CONFIGURATION 1 OF 1

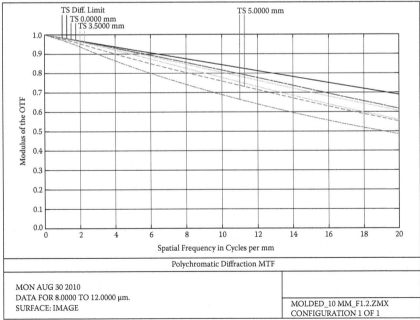

Polychromatic Diffraction MTF

MON AUG 30 2010
DATA FOR 8.0000 TO 12.0000 μm.
SURFACE: IMAGE

MOLDED_10 MM_F1.2.ZMX
CONFIGURATION 1 OF 1

FIGURE 6.7 (continued)

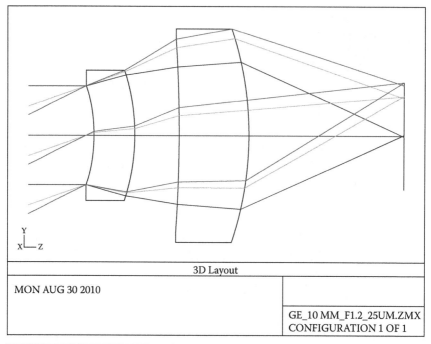

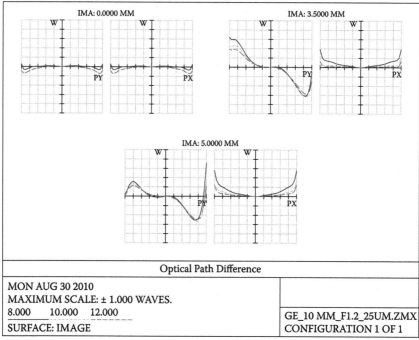

FIGURE 6.8
A traditional two-element germanium retrofocus lens.

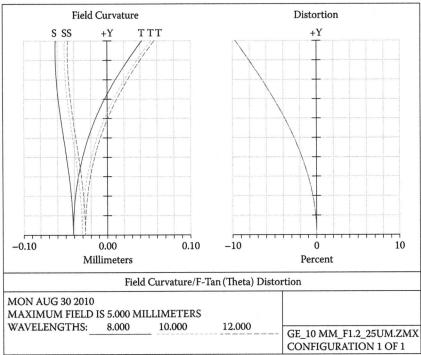

Field Curvature/F-Tan (Theta) Distortion

MON AUG 30 2010
MAXIMUM FIELD IS 5.000 MILLIMETERS
WAVELENGTHS: 8.000   10.000   12.000

GE_10 MM_F1.2_25UM.ZMX
CONFIGURATION 1 OF 1

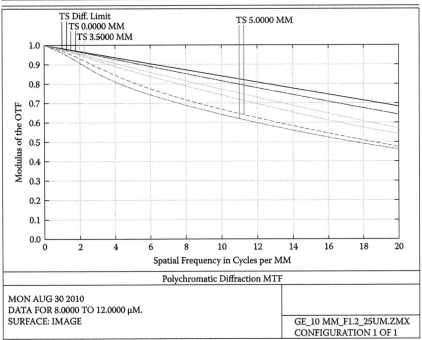

Polychromatic Diffraction MTF

MON AUG 30 2010
DATA FOR 8.0000 TO 12.0000 μM.
SURFACE: IMAGE

GE_10 MM_F1.2_25UM.ZMX
CONFIGURATION 1 OF 1

FIGURE 6.8 (continued)

are mounted in a rigid aluminum housing, the focus of the molded infrared lens will shift by ±0.055 mm over this temperature range. However, the thermal focus shift of the traditional lens is nearly three times as much, ±0.160 mm. For an F/1.2 lens operating in the 8 to 12 μm waveband, the diffraction-limited quarter-wave depth of focus is approximately ±0.024 mm, which tells us that the molded lens will suffer degraded but potentially usable image quality at the temperature extremes. However, the traditional germanium lens would sustain a significant performance loss and require refocus or mechanical athermalization to recover performance.

Using a retrofocus layout a molded infrared lens with a short focal length can provide a back focal distance large enough to accommodate the camera package. With heavy use of aspheric surfaces and a diffractive element a molded lens can provide image quality that rivals that of traditionally fabricated infrared lenses. For fixed-focus lenses that must perform at temperature extremes, the molded lens might be preferred since its thermal focus shift is much less than that of an equivalent germanium lens design.

6.4.4.2 The Molded Infrared Petzval Lens

As the focal length of an LWIR lens becomes longer, the back focal distance requirement no longer drives the lens form; however, minimizing the overall size of the lens can become a pressing requirement. In this situation a Petzval lens layout offers an effective solution that is readily adopted for molded infrared lenses. Traditionally, a Petzval lens operating in the visible waveband provides good image quality in the regime where the horizontal field of view is on the order of 10° to 20° and the lens is working in the vicinity of F/1.5.[21] This lends itself nicely to uncooled LWIR systems with focal lengths in the 50 to 100 mm range. Molded chalcogenide glasses can also be used to produce a Petzval lens shorter than 50 mm, but as the field of view increases, the rear lens might need to be split to provide more degrees of freedom for aberration correction.

A classical Petzval lens divides the focal power between two positive powered lens groups. To obtain a system effective focal length of f, the lens groups are separated by a distance f and the focal length of the front and rear groups is $2f$ and f, respectively. A back focal distance of $f/2$ is produced. The aperture stop is placed at the front lens and the two lens groups are individually achromatized.

A modern infrared Petzval lens can utilize low-dispersion materials and aspheric surfaces to reduce the number of elements. In the LWIR, aspheric surfaces and the extremely low dispersion of germanium allow the lens to function using two singlet lenses rather than achromatized groups. A lens constructed from molded materials readily lends itself to this layout. Higher dispersion in the moldable infrared chalcogenides requires a diffractive surface for color correction. As the focal length increases and the field of view narrows, axial color becomes more of an issue than lateral color. Thus, the diffractive surface is most efficient when placed near the aperture stop (on the first lens). Power is then shifted to the first lens so that less axial and lateral color are introduced by the weaker second element, which also drives the lenses closer together. This can cause the focal length and spacing of the elements in a molded chalcogenide Petzval lens to diverge slightly from the classical layout, but the analogy is still useful.

For fast lenses heavy usage of aspheres might be required to achieve diffraction-limited performance for frequencies below the detector's limiting resolution. As a cautionary note, when high-order aspheres are placed on both sides of a lens, the designer should be careful to ensure that the aspheric components are providing useful aberration correction. In some situations these lenses can converge to a highly bowed or rippled shape with inflections in the local surface curvature. While sometimes useful, this might unnecessarily complicate the fabrication and test of the lens. Adequate sampling of the pupil and field should also be used to ensure highly aspheric lenses don't introduce poorly optimized zones within the FOV. As the focal length of a molded infrared Petzval lens increases from 50 mm to 100 mm (and beyond), fewer aspheric surfaces are likely required because the field of view of the lens decreases.

Figure 6.9 provides an example of a moldable F/1.2 50 mm Petzval lens. The lens is designed to operate with an uncooled 640 × 480 detector with 25-micron pixels, producing a horizontal FOV of 18.2°. The optimization merit function was selected to maximize the performance below the Nyquist frequency of 20 cycles/mm. A kinoform diffractive surface with fifteen zones is placed on the concave side of the first element to correct axial color. Image quality was driven to near the diffraction limit by aspherizing every surface. Field curvature is minimized by using higher-index chalcogenides because both lenses have a positive focal length. In this example, a Ge-Sb-Se chalcogenide glass (IG5, N @ 10 μm = 2.604) was used because it provides a refractive index slightly higher than chalcogenides from the alternative Ge-As-Se family.

For comparison, a traditional germanium Petzval lens is shown in Figure 6.10. The lens very closely follows the classical Petzval layout. Equivalent image quality to the molded lens is produced with three low-order aspheric surfaces. A diffractive surface is not required for color correction, as indicated by the small amount of axial color observable in the wavefront aberration plot in the upper right.

As with the retrofocus example, the most significant functional difference between the molded and traditional lenses is the thermal performance. The molded lens will have a higher operational temperature limit since it does not suffer from germanium's increasing absorption with temperature. More importantly, the shift in focal position with temperature becomes significant since its magnitude scales with a lens's focal length. The back focal distance of the germanium system shifts by ±0.320 mm over the −40 to +80°C temperature range. By comparison, the much smaller thermoptic coefficient of the molded lenses shifts the focus by only ±0.184 mm. The difference is nearly

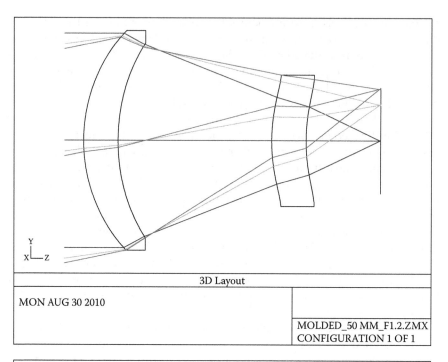

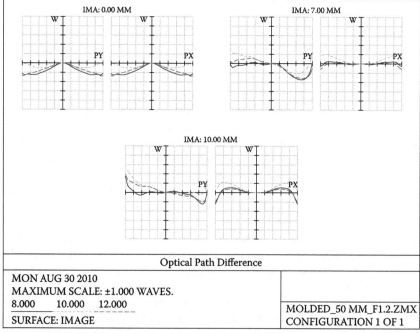

FIGURE 6.9
An F/1.2 molded chalcogenide Petzval lens with a 50 mm focal length.

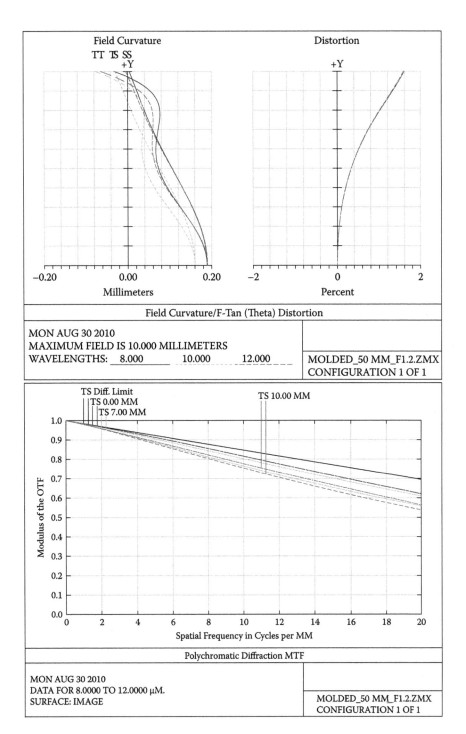

FIGURE 6.9 (continued)

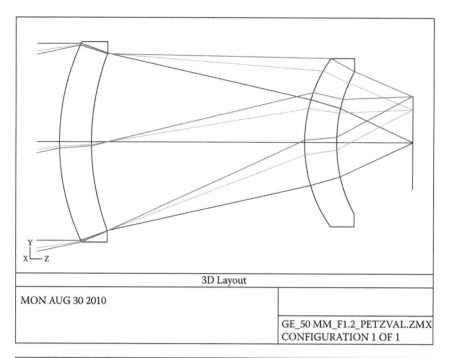

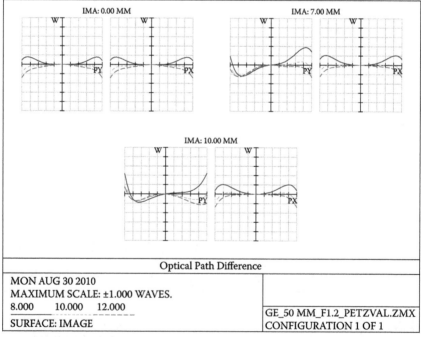

FIGURE 6.10
A classical germanium Petzval lens.

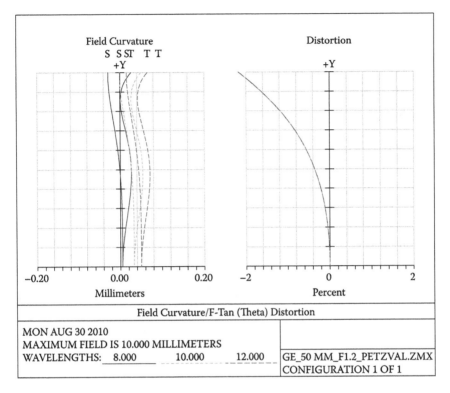

FIGURE 6.10 (continued)

a factor of two, which can play a major role in the mechanical design of a system that must be passively athermalized.

6.5 Concluding Remarks

Many classical infrared design forms can be constructed using molded optics. Usage of molded diffractive surfaces can allow the lenses to function like achromatic doublets, and molded aspheres can be used to correct aberration and achieve diffraction-limited performance with fast F/numbers. The application of molding technology to infrared lenses has shown great potential in recent years. A growing market for SWIR sensors provides a new platform for using molded glasses and plastics. Processes developed for molding plastics and visible glasses have been transferred to infrared chalcogenides to accurately produce complex surfaces with tolerances rivaling those achievable by conventional fabrication. Molded chalcogenide lenses can offer a significant advantage in high-volume applications since

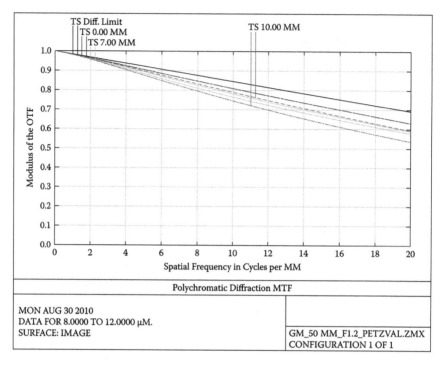

FIGURE 6.10 (continued)

the cost of their manufacture and material can be significantly lower than that for traditional infrared lenses. Furthermore, enhancements to the thermal performance of a lens can be obtained by using chalcogenide glasses as a substitute material for systems that typically rely on germanium optics.

Molding infrared lenses is still a relatively new process. In just the past few years great strides have been made toward pushing the upper limit of moldable lens size and increasing the commercial availably of durable high-quality antireflection coatings. The ability to tailor the composition of chalcogenide glasses to optimize cost, glass transition temperature, and optical or thermal properties holds great promise. As demand for molded chalcogenide lenses increases, a greater variety of material and coating options is anticipated, as well as continued reduction in fabrication costs.

References

1. Shepard, R. H., Sparrold, S. 2008. Material selection for color correction in the short-wave infrared. *Proc. of SPIE* 7060:70600E.
2. Amorphous Materials, Inc. 2010. http://www.amorphousmaterials.com
3. Umicore Electro-Optic Materials. 2010. http://eom.umicore.com
4. Schott North America for VITRON Spezialwerkstoffe GmbH. 2010. www.us.schott.com/advanced_optics
5. Crastes, A., et al. 2003. Low cost uncooled IRFPA and molded IR lenses for enhanced driver vision. *Proc. of SPIE* 5251, pp. 272–279.
6. Guimond, Y., Franks, J., Bellec, Y. 2004. Comparison of performances between GASIR moulded optics and existing IR optics. *Proc. of SPIE* 5406:114.
7. U.S. Geological Survey. 1998-2008. *Minerals yearbook: Metals & minerals*. Vol. I. U.S. Geological Survey.
8. Curatu, G., et al. 2006. Using molded chalcogenide glass technology to reduce cost in a compact wide-angle thermal imaging lens. *Proc. of SPIE* Vol. 6206, 62062M.
9. Zhang, X. H., Guimond, Y., Bellec, Y. 2003. Production of complex chalcogenide glass optics by molding for thermal imaging. *J. Non-Crystalline Solids* 326–27:519–23.
10. Graham, A. et al. 2003. Low cost infrared glass for IR imaging applications. *SPIE* 508:216–23.
11. Cordier, C., Lonnoy, J. 2004. IR low cost molded optics. *SPIE* 5252:92–102.
12. Hilton Sr., A. R., et al. 2008. Amorphous materials molded IR lens progress report. *Proc. of SPIE* 6940:69400Q.
13. Hilton Sr., A. R. 2005. Development of chalcogenide glasses as optical materials for infrared systems. *Proc. of SPIE* 5786:258–61.
14. Hilton Sr., A. R. 2010. *Chalcogenide glasses for infrared optics*. New York: McGraw-Hill.
15. Bacchus, J. M. 2004. Using new optical materials and DOE in low-cost lenses for uncooled IR cameras, optical design and engineering. *Proc. of SPIE* 5249:425–32.
16. DRS Technologies. 2010. U6000 product brochure. www.drsinfrared.com
17. FLIR Systems. 2010. Photon 640 product brochure. www.flir.com
18. Raytheon Company. 2010. U640 product brochure. www.raytheon.com
19. BAE Systems. 2010. SC550 product brochure. www.baesystems.com
20. Curatu, G. 2008. Design and fabrication of low-cost thermal imaging optics using precision chalcogenide glass molding. *Proc. of SPIE* Vol. 7060, 706008.
21. Smith, W. J. 2005. *Modern lens design*, 42. 2nd ed. Bellingham, WA: SPIE Press.
22. U.S. Geological Survey. 2010. Commodity statistics and information. http://minerals.usgs.gov/minerals/pubs/commodity

7

Testing Molded Optics

Michael Schaub and Eric Fest

CONTENTS

7.1 Introduction

There are two main categories of testing with respect to molded optics. These are the testing of molded optic components and the testing of systems incorporating them. Based on these two categories, we discuss some of the methods and equipment normally seen in the testing of molded optics. Many of the tests performed on molded optics are also performed on standard optics and are covered in other texts. We therefore focus on testing that is more specific to molded optics, such as aspheric surface form measurement and stray light evaluation.

7.2 Molded Optical Components

The primary characteristics of molded optical components that are tested are their surface form and surface centration. There are multiple methods of testing each of these, which are discussed below. Often, the test method is selected based on the available test equipment.

7.2.1 Surface Form

7.2.1.1 Interferometry

Testing of optical surfaces is performed in order to verify that each surface has the proper shape. As discussed earlier, molded optical surfaces can take a variety of forms. These include planar, spherical, or aspheric shapes, as well as surfaces with structures such as diffractives, lenslet arrays, or Fresnel facets. The molded optical surface may be a combination of types, such as a diffractive on an aspheric base, or it may consist of several sections containing different form types.

Measurement of surface form is most often accomplished through the use of either interferometers or surface contact profilers. We begin with a discussion of interferometry, a well-established method of optical testing that has been described and documented in a number of texts.[1,2] Interferometry relies upon the principle of interference of light, that is, the interaction of light waves when brought together. This interaction, when viewed by a camera or the eye, results in a pattern of bright and dark bands called fringes. Properly interpreted, the fringe pattern (referred to as an interferogram) provides information regarding the form of the surface or, more directly, the departure of the surface form from a known reference form.

In most cases of interferometric testing the interference occurs between a test beam, which represents the surface under test, and a reference beam, which represents the ideal shape that the surface is being compared to. The reference beam usually has a planar or spherical wavefront, as these are the shapes that can most easily be generated. In the case of surface testing (as opposed to testing the beam transmitted by the element), the test beam wavefront shape normally results from reflection off of the test surface, while the reference beam is generated by reflection from a precision surface within the test equipment.

Figure 7.1 shows a simple surface form testing case, the interferometric measurement of a convex spherical surface. On the left of the figure is the interferometer, while on the right is the surface being tested. The interferometer consists of a laser source, optical elements including the reference sphere, as well as a camera system to capture the interferogram. In this particular setup, we are determining departure of the test surface from true sphericity. The spherical reference wavefront is generated by the reference

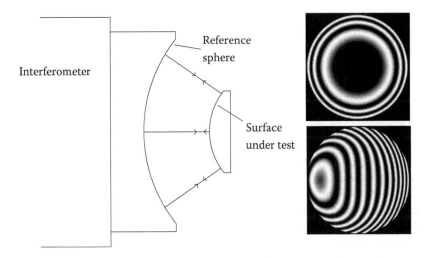

FIGURE 7.1
Interferometric testing of a convex spherical surface (left). Interferogram showing fringes for sag error at edge of surface (right. top). Interferogram showing fringes for sag error at edge of surface with addition of tilt (right, bottom).

sphere, from reflection off of its final optical surface. This reflected wavefront (the reference beam) travels back into the interferometer and will interfere with the test beam to provide the fringe pattern. The reference sphere also produces and transmits the test beam, itself spherical, which will reflect off of the test surface and return to the interferometer. To measure sphericity, the convex surface is positioned such that the converging spherical wavefront exiting the reference sphere is focused at the surface's center of curvature. Any imperfections in the surface under test will show up in the reflected test beam wavefront. For instance, if the test surface is not perfectly spherical, but sags at the edge, the reflected beam will also show this nonspherical trait. When interfered with the reference beam, a set of fringes indicating this departure will be seen. Each fringe corresponds to one-half of a wave of light (at the wavelength of the interferometer laser) of height departure from the reference surface, in this case a sphere. In this way, the fringes act as a topographic map, showing the departure of the test surface from the reference shape.

If the test surface were perfectly spherical, and perfectly aligned with the interferometer, the fringe pattern seen would not appear as much of a pattern at all, but as a single, uniform color. Often, it is not possible (or desired) to perfectly align the surface under test to the interferometer. If there is a small amount of tilt between the test surface and interferometer, tilt fringes will be added to the interferogram. Tilt fringes are a set of parallel, evenly spaced fringes, whose orientation displays the direction of the tilt angle between the test surface and interferometer. The number of fringes is related to the magnitude of the tilt angle. In many cases, the addition of tilt fringes can make visual interpretation of the fringe pattern easier. If the surface under test had a bump near the center, we would see a "wiggle" of the fringes in the area of the interferogram corresponding to the area of the surface containing the bump. Similarly, if the surface had slightly different radii of curvature in orthogonal axes, we would see this in the fringe pattern.

The final interpretation of an interferogram is usually performed by a computer program associated with the interferometer. The resulting output can be compared to the tolerance on the part specification, and a pass/fail determination made. Surface departure maps can be generated, if desired, and placed onto the surfaces in the optical design code. This allows prediction of the impact of the actual surfaces on performance of the system.

It should be noted that the direction of an error cannot be determined from a single interferogram. We previously mentioned that a surface with a bump at its center would result in a wiggle of the fringes at the center of the interferogram. Looking at the wiggle, however, we cannot tell if the error is a bump or a hole. The direction of the error can be determined by evaluating the motion of the fringes when a slight displacement is applied to the reference sphere or object under test. Previously, the person performing the measurement would gently push on the mount holding the optic under test to make this determination. Today this is unnecessary, as most commercial interferometers take multiple interferograms with precise motion applied to the reference sphere using piezoelectric transducers. These multiple interferograms are run through an (usually proprietary) algorithm and the results displayed on the monitor. The use of multiple interferograms and appropriate algorithms can also help to reduce systematic errors in the measurement equipment.

In this example, if we want to determine the radius of curvature of the spherical surface, we move the surface away from the interferometer until the test beam focuses on the surface itself. This position is known as cat's eye for its similarity to the bright reflection seen from the retina of a cat. The distance that the surface is moved, from the test beam being focused on the center of curvature to the test beam being focused on the surface itself, is the radius of curvature of the surface. This motion is usually accomplished with an air-bearing or precision slide and measured with an encoder or distance-measuring interferometer. This allows the value of the radius to be measured to within microns.

While this example illustrates the testing of a convex spherical surface, setups to test plano or concave spherical surfaces can easily be considered. Most commercial interferometer producers sell a range of reference sphere sizes and F/#s, along with auxiliary mounting equipment, to allow the testing of a wide range of spherical or plano surfaces. However, as was previously stated, one of the advantages of molded optics is the ability to produce and utilize aspheric surfaces. We now discuss the testing of aspheric surface form using interferometric means. We mentioned that planar or spherical reference wavefronts are usually output from interferometers, because of the relative ease of their creation. When testing an aspheric surface, it is generally not useful to compare it to a spherical surface, unless the aspheric departure from sphericity is small. This is easily understood by considering our earlier statement that each fringe in the interferogram corresponds to a height departure of one-half of a wavelength. The most common laser source for interferometers is the HeNe laser, with a wavelength of 0.6328 microns. Therefore, if the aspheric surface has only a micron (approximately three half-wavelengths) sag departure from a sphere, only a few fringes will be generated in the interferogram. As the aspheric departure increases the number of fringes will also increase, until there are so many fringes that they can no longer be resolved on the camera within the interferometer. Once this condition is reached, false readings will occur.

In order to properly test the aspheric surface, the density of fringes in the interferogram needs to be controlled. This can be accomplished through the use of a null lens. A null lens is an auxiliary optic that converts the flat or spherical wavefront exiting the interferometer into a beam with a wavefront shape that closely or exactly matches the shape of the aspheric surface under test. If the null lens made a wavefront shape exactly matching the shape of the surface under test, only a uniform color interferogram, known as a null, would be produced (under perfect alignment). Hence, the auxiliary optic is termed a null lens. Figure 7.2 shows a test setup using a diffractive null lens to test a convex, aspheric surface. The reference beam is generated by reflection from the reference flat. The (planar wavefront) output from the interferometer enters the null lens and is converted into a shape matching the desired aspheric surface. After reflecting from the aspheric surface, the wavefront passes back through the null lens and is converted back into a

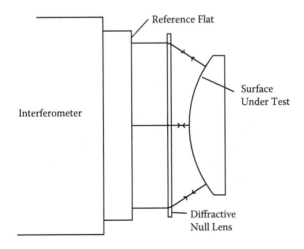

FIGURE 7.2
Interferometric testing of a convex aspheric surface using a diffractive null lens.

planar wavefront (if the aspheric surface is perfectly made). Imperfections in the surface under test will result in imperfections in the planar-like wavefront entering the interferometer. These imperfections will be seen in the resulting fringe pattern, allowing the magnitude of the error in the surface to be calculated.

Null lenses can be constructed in a variety of ways. The diffractive null lens shown in the figure consists of a holographic pattern written onto a glass plate. Alternatively, null lenses can produce the desired wavefront shape through refractive, reflective, or combinations of refractive/reflective/diffractive means. The design of the null lens can be performed using the same optical design and analysis software that was used to create the aspheric surface prescription; null lenses are often designed by the same person that specified the asphere. However, they may also be designed by the manufacturer, who will be using the null during production of the aspheric surface. One of the keys to designing a null lens is to ensure that it will work properly over the allowed tolerance range of the surface to be tested. Using modern optical design software, it is generally quite easy to design a null lens to test the nominal aspheric surface. In reality, the surface produced will have some variation from the nominal and the null lens needs to perform under this range of variation. A tolerance analysis on the test setup, which includes the tolerances on both the null lens and the aspheric surface, is strongly recommended to prevent development of an inadequate null lens. The necessary alignment resolution for the test setup can also be determined from this analysis.

Interferometric measurement of the radius of curvature of an aspheric surface is not as simple as measuring the radius of a spherical surface. When using a null lens, the radius of curvature of the aspheric surface can be determined by evaluating the residual power (curvature) in the interferogram. Achieving an accurate measurement requires precise positioning of the null lens and the surface being tested. This may be accomplished using well-characterized fixturing or metering rods. Small errors in the physical setup may result in significant measurement error, again emphasizing the need to perform a tolerance analysis on the test arrangement itself.

A more recent development in the interferometric measurement of aspheric surfaces is the use of "stitching" interferometers.[3] In this method, a number of subaperture interferograms are taken, without a null, and are stitched together to create an interferogram of the entire surface. The size of the subaperture interferograms, along with the relative positions of the reference sphere and surface being tested, are controlled such that a measurable fringe density is maintained.

7.2.1.2 Profilometry

The second and potentially more commonly used method of measuring the surface form of molded optics is through the use of a surface contact profiler.

The reasons for this are the profiler's relative ease of use and adaptability. Unlike interferometric measurement of aspheric surfaces, which generally requires an individual null lens for each surface shape, a surface profiler does not need auxiliary equipment to measure a wide range of surface shapes. Additionally, the newer surface profiler models can self-align to the surface under test, reducing the need for a highly trained test technician.

Surface profilometers measure surface form by scanning a stylus across the optical surface. To obtain a surface form profile, the stylus is first placed in contact with the surface to be measured. The position of the stylus is adjusted so that the scan will run through the center of the optical surface (assuming a rotationally symmetric surface). The stylus is next moved near the edge of the part, typically slightly outside the surface's clear aperture. Maintaining contact with the surface, it is then translated across the diameter of the part. During this motion the height of the stylus as a function of lateral position is recorded. The height vs. position data (the surface profile data) can be analyzed and displayed in a number of ways. Most often, the data are compared to the desired surface, with a plot of the difference between the desired and measured profile displayed. Often, a best-fit radius is first calculated, so that any radius tolerance error can be evaluated and removed. Note that in this way the radius value of the surface, as well as the aspheric form error, is obtained in the measurement.

The main drawback to measuring optical surfaces with a surface contact profiler is that each measurement produces data only along a single line traversing the surface. Errors not on the path traveled by the stylus are not revealed. Thus, it is generally appropriate to request that more than one profile be measured on a given surface. For molded plastic optics it is common to specify two profile measurements to be made: one aligned with the injection gate, and another orthogonal to this trace. This will show any asymmetry due to the flow of plastic into the mold cavity.

While the stylus does contact the surface, it usually does not damage it, at least to an extent that would cause failure of the part. Small indentations on the part surface may be seen, especially in plastic optics, under significant magnification. The pressure applied by the stylus is normally adjusted to minimize the impact to the surface, while still maintaining sufficient contact to ensure a valid measurement.

7.2.2 Surface Position

As seen earlier with the cell phone camera lens example, surface position can have a large impact on the ultimate performance of a system using molded optical components. Surface position measurements are used to verify the centration or tilt of the molded optical surface. Surface position is normally determined either through a mechanical or optical run-out measurement, or through inspection of features on the optical surface.

7.2.2.1 Mechanical Run-Out

When a nonplanar optical surface is decentered, it results in a difference in surface sag, at a given radial distance, from the original centered axis (usually referenced from the part diameter). The magnitude of this sag difference will depend upon the amount of surface decentration. Thus, the sag difference provides information on the position (centration) of the surface. Measurement of the sag difference of a surface is typically performed by placing a dial indicator in contact with the optical surface and rotating the component about its axis. The full motion of the indicator (plus and minus) is referred to as the run-out of the part, and can be converted into the decenter or tilt of the surface. Care should be exercised in the measurement and conversion, particularly when testing steeply curved surfaces. When specifying a run-out, it is generally assumed that the dial indicator will be placed normal to the surface at the location the measurement is made. If the indicator is instead positioned such that it moves vertically, instead of normally to the surface, there will be a corresponding error in the measurement. Normal-to-surface motion (or lack thereof) needs to be accounted for in the specification of run-out and its conversion to decenter.

For spherical surfaces, tilt and decenter have the same effect on the surface. This results from the fact that the sphere has no defined axis of symmetry, only a radius value. An aspheric surface, on the other hand, has a defined axis. Because of this, a mechanical run-out measurement cannot differentiate between tilt and decenter of the surface. In fact, it is theoretically possible for tilt and decenter errors to mask each other during the mechanical run-out measurement. In general, however, molded optic components rarely suffer from the tilt of their individual optical surfaces. This is a result of the methods, such as diamond turning, which are used to produce the molds they are replicated from. Thus, mechanical run-out measurements are usually applied to surface decentration of molded optical components.

Tilt of the surfaces of molded optical components can often be evaluated based on secondary features of the part. For instance, when diamond turning the optic insert for a convex molded optic, there is often a small annular ring placed around the optical surface. This annulus prevents the optic insert from having a sharp edge, which could be easily damaged when the insert in placed in the mold. Because the annular flat and the optical surface are machined in the same diamond turning setup, they are essentially perfectly aligned to one another in regard to their tilt. Thus, measurement of the tilt of the flat annular ring with respect to the mounting flange on a molded optic can provide equivalent information on the tilt of the optical surface.

7.2.2.2 Optical Run-Out

Similar to the rotation of a decentered surface causing motion of a dial indicator placed against it, rotation of a decentered optical surface will cause motion of a beam reflected off it, referred to as optical run-out. Optical run-out measurements, like mechanical run-out measurements, can provide information on the position of the optical surface. Optical run-out measurements are performed by observing the full range of motion of an image as the component is rotated about its optical axis. The observed image may result either from transmission through the element or from reflection off of one of its surfaces. Observing the motion of the reflected image provides information on the individual surface, while observation of the motion of the transmitted image provides information on the element as a whole. Transmitted image motion can also be used to align two elements, as in the production of a cemented glass doublet. As with mechanical run-out measurements, the designer needs to convert the image motion to an equivalent decentration or tilt. This conversion can be performed in the optical design software, either manually or through program features.

7.2.2.3 Surface Features

Surface features on molded optics can also be used to determine information on surface position. For molded plastic optics in particular, there is often a tiny spot at the center of the optical surface that is visible under high magnification. This feature is produced as a result of the diamond turning of the optic insert. The spot on the insert is then transferred to the optic during the molding process. Because the feature is produced during the diamond turning process, its location is precisely centered on the optic surface. Similar to performing a run-out measurement, the motion of the diamond turning feature can be observed as the part is rotated, giving a direct indication of the surface decentration. Measurement of the features on the front and rear surfaces of an element provides the surface-to-surface registration of the part. If the second surface is observed through the first surface, which is usually the preferred method, the magnification of the front surface needs to be accounted for. This method of measuring surface centration, based on diamond turning features, has been independently developed by several organizations and has been reported on.[4]

In addition to this central diamond turning feature, other features on the part may also be used. We mentioned the use of the annular flat for tilt measurements above. The transition of the optical surface to this flat, which is again diamond turned and well located, can be used as a reference feature as well. In addition to visual inspection and measurement, features such as these can be used in evaluation of the molded optics by automated inspection systems.

7.2.3 Diffractive Surface Features

We discussed earlier the impact that "dead zones," the rounding of the top and bottom of diffractive features, can have on system performance. We also noted that the height of the diffractive features determines the wavelength that has (theoretically) perfect diffraction efficiency. The measurement of diffractive surface features is usually performed using a white light interferometer, which externally appears similar to a microscope, with a focusing objective placed above the surface being tested. The objective is mated to a detector array inside the interferometer, which is used to image a two-dimensional area of the surface under test. As with the interferometer discussed above, a white light interferometer relies upon the interference of light waves. In this instrument, however, the laser source is replaced with a (possibly filtered) white light source. This results in multiple fringe patterns being generated, with a pattern created for each wavelength in the source. Only at one position, when all of the fringe patterns are aligned, is a high-contrast fringe pattern obtained. This occurs when the light is focused directly on the optical surface. Thus, by scanning the focusing objective and noting the locations when the fringe pattern on each detector pixel has highest contrast, a height map of the surface is generated. Using this height map the depth of the diffractive features, as well as any rounding of their edges, can be evaluated.

Surface profilers can also be used to evaluate diffractive microstructures. These are generally more limited in their capabilities, though, due to the finite size of the stylus. The measurement displayed by the surface profiler is actually the convolution of the stylus tip shape with the surface feature measured. It is possible to perform a deconvolution within the software of the profiler. However, the results should be carefully evaluated, particularly for very fine diffractive features.

7.2.4 Materials

While the material for a molded optical component is usually specified on the part drawing, the actual material used to produce any given component is normally not subjected to any testing by the molder. Instead, the molder relies upon the material manufacturer to provide the correct material to it. Once the material is received by the molder, it relies upon its internal handling procedures to ensure that the material is properly identified, segregated, and stored. The manufacturer of the material often supplies a certificate of conformance to the molder, who can provide a copy to the customer if desired.

If material testing is needed, for instance, for measurement of the refractive index of a material or for coating testing, test articles can be fabricated. In the case of the refractive index, the test article produced is usually a prism. The refractive index of the prism can be determined using standard methods such as the minimum deviation method. Refractive index measurement as a function of wavelength and temperature is available through commercial services.[5]

When performing material property measurements, it is preferred that the test article represent the actual part as closely as possible. Ideally, this means that the test article is made from the same material lot and processed in the same manner as the parts it represents. Because the molding process parameters can affect the material characteristics, such as the refractive index, the test articles should be created under as similar conditions as possible. In some cases this is not possible, due to size differences between the part and test article. In other cases, it may not be economically feasible. It may be possible, however, if the parts are sufficiently large, to machine the test articles from the actual parts. In this way, the material will be exposed to the exact molding process, and any differences between the part and test article will only be due to the secondary machining process.

7.3 Systems Using Molded Optical Components

Systems using molded optical components are subjected to many of the same tests as standard optical systems. These tests include transmitted wavefront, modulation transfer function (MTF) performance, and evaluation of first-order properties such as focal length. We do not discuss these here, but concentrate instead on stray light testing and evaluation of diffraction efficiency.

7.3.1 Stray Light Testing

As discussed in Chapter 3, the last step in the stray light engineering process is the hardware test. The purpose of this test is usually to determine if the system meets its stray light requirements and to validate, as much as possible, the results of the stray light analysis. For systems with very low stray light levels, validating the model may be difficult or impossible because the testing procedure may not be sensitive enough. There are a number of methods of testing stray light, and some are better suited to testing a particular type of stray light requirement than others. Two types of stray light tests will be discussed here: solar tests and veiling glare tests.

7.3.1.1 Solar Tests

The purpose of a solar test is to measure the amount of stray light on the focal plane from illumination by the sun or a sun-like source. Since the sun is usually the most commonly encountered source of stray light, solar stray light requirements and tests are very common. A solar test is often an absolute test, in that the goal of the testing procedure is to quantify the irradiance at the focal plane of the sensor due to stray light. In order to perform an absolute test, it is usually necessary to first calibrate the sensor, which allows the sensor output to be correlated to a known irradiance level at the focal plane. This can be done by using the sensor to look at a source of known brightness, often a calibrated lamp (for the visible) or blackbody (for the IR) whose radiance has been measured very accurately.[6,7] The source is put in the FOV of the sensor and the output of the sensor is recorded. This output can then be used to compute a scale factor S = (sensor output level)/(focal plane irradiance), where the sensor output level is a measure of the focal plane output (for digital cameras, usually a grayscale value), and the focal plane irradiance is computed using Equation 3.20. Other calibration sources include integrating spheres and diffuser plates.[8]

Once the sensor is calibrated, the solar test can proceed. The simplest way to perform this test is to just use the sun itself to illuminate the sensor. It is usually necessary to know the azimuth and elevation angles that the sun is at relative to the optic axis. This can be done by consulting a solar azimuth and elevation calculator[9] to determine the range of solar angles available for a particular location and time period, and then aligning the optic axis with a compass to locate the sun at the desired off-axis angles. It is usually easiest to do this when the sensor is mounted on a tripod. The output on the sensor at each solar angle can then be scaled using the scale factor S to determine the stray light irradiance on the focal plane, which can then be compared to the requirement and the model results.

There are a number of difficulties in using direct sunlight to perform the test. The first is that the range of solar angles is limited by the location, time of day, and ability to precisely orient the optic axis of the sensor. Another is that the sun moves, which may be a problem in testing systems requiring long exposures. A solution to these problems is to use a heliostat,[10] which is a device designed to reflect the sun in such a way as to keep its angle of incidence on the sensor constant. Heliostats work by tracking the sun and using the tracking information to move a mirror to keep the angle of the mirror reflection constant on the sensor, as shown in Figure 7.3. This allows for more control over the range of angles incident on the sensor and allows the image of the sun to stay constant.

Solar tests performed using the sun itself are very common; however, they have some drawbacks. One is that the solar radiance is a strong function of weather conditions. This introduces variability into the test; the radiance of the sun can vary day to day and even hour to hour, and this can make determining the observed solar radiance and comparing the results of different tests difficult. Another drawback is that solar tests usually must be performed outside, and thus the system and its test equipment may be exposed to dust and other contaminants. One way to avoid this problem is to use a laboratory solar simulator, either a lamp[11] (for visible systems) or blackbody[2] (for IR systems). These sources are collimated by an optical system, shown schematically in Figure 7.4. The apparent angular extent of the simulated solar source can be different than the angular extent of the sun (which is about 0.53°), and the effect of this difference should be evaluated before using a simulator.

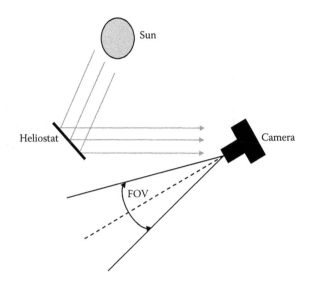

FIGURE 7.3
Solar stray light testing using a heliostat.

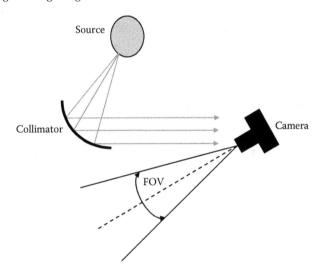

FIGURE 7.4
Solar stray light testing using a laboratory solar simulator.

7.3.1.2 Veiling Glare Test

The veiling glare index (VGI) is a means of quantifying the stray light performance of an optical system (usually only systems operating at visible wavelengths), and was discussed in Section 3.5.1.2. VGI is equal to the ratio of flux at the image plane due to uniform illumination from outside the sensor FOV, divided by the flux at the image plane due to uniform illumination

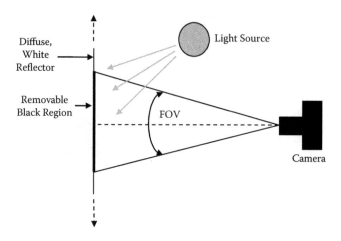

FIGURE 7.5
Veiling glare test arrangement.

both inside and outside the FOV. VGI is tested using a veiling glare test, which is usually performed using a broad, white diffuse reflector (such as a white wall or screen) with a small, removable black region in the center, as shown in Figure 7.5. The reflector is illuminated using a light source that is large enough to uniformly illuminate it. The broader the diffuse reflector, the more accurate the test will be. The black region is sized to exactly subtend the sensor FOV (which is rectangular for most digital camera systems). The measurement is performed by pointing the sensor at the diffuse source without the black region and sampling the sensor output (in digital cameras, this is just taking an image), and then positioning the black region so that it just fills the FOV and then taking another sample. The values of Φ_{out} and Φ_{in} in Equation 3.31 can then be computed by summing the magnitude of each sensor output sample over the FOV for each sample taken (Φ_{out} for the first, Φ_{in} for the second). For digital cameras, this corresponds to summing the grayscale values of all of the pixels in each image. This test is relative, and therefore no calibration is necessary, although care should be taken not to saturate the sensor output.

7.3.2 Diffraction Efficiency

The use of diffractive surfaces and issues with their diffraction efficiency have been discussed earlier in the text. Testing of diffraction efficiency can be performed at either or both the component and the system level; testing at the component level may aid in production of the element, while testing at a higher assembly level determines the ultimate impact of the surface on system performance. Diffraction efficiency testing can be performed in a qualitative or quantitative manner. Qualitative testing typically involves visual evaluation of the image produced by the system or component, ideally under

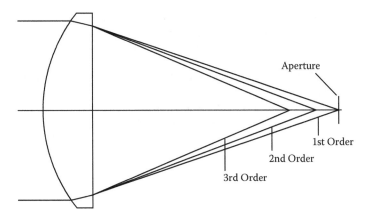

FIGURE 7.6
Diffractive efficiency testing.

as-used conditions. Quantitative tests can be performed to determine the actual diffraction efficiency or to evaluate the ability of the system to perform a given task.

A simple test arrangement for evaluating diffraction efficiency is shown in Figure 7.6. This method involves the use of a small aperture to isolate the light in a given diffraction order. Moving the aperture along the axis of the system, multiple diffractive order foci are evaluated. Taking the ratio of the light in the design order to the sum of light in all the orders yields the diffraction efficiency. This method assumes that the foci are separated far enough, or that the aperture is small enough, such that out-of-focus light from the other orders (ones not currently being evaluated) does not provide a significant contribution (error) to the light in the measured order. This measurement can also be performed for an off-axis input beam, where the diffracted orders may be laterally separated. The use of filtered multiwavelength sources, or multiple lasers, allows evaluation of the diffraction efficiency as a function of wavelength as well as order.

References

1. Malacara, D. 2007. *Optical shop testing*. Hoboken, NJ: John Wiley & Sons.
2. Goodwin, E. P., and J. C. Wyant. 2006. *Field guide to interferometric testing*. Bellingham, WA: SPIE.
3. Haensel, T., A. Nickel, and A. Schindler. 2001. Stitching interferometry of aspherical surfaces. *Proc. SPIE* 4449:265–75.
4. Olson, C. 2005. Precision centration measurements on injection-molded lenses for digitial imaging. *Proc. SPIE* 5872:587204.

5. Ohara Corp. http://www.oharacorp.com/measure.html.
6. Gamma Scientific. http://www.gamma-sci.com.
7. Infrared Systems Development Corp. http://infraredsystems.com.
8. LabSphere, Inc. http://www.labsphere.com.
9. NOAA ESRL. Solar position calculator. http://www.srrb.noaa.gov/highlights/sunrise/azel.html.
10. Heliotrack.com. http://www.heliotrack.com/
11. Newport Corp. http://www.newport.com.

Index

Milton Keynes UK
Ingram Content Group UK Ltd.
UKHW040109071024
449327UK00019B/927